SpringerBriefs in Applied Sciences and Technology

More information about this series at http://www.springer.com/series/8884

Christopher W. Foster · Rashid O. Kadara
Craig E. Banks

Screen-Printing Electrochemical Architectures

Springer

Christopher W. Foster
Faculty of Science and Engineering
Manchester Metropolitan University
Manchester
UK

Craig E. Banks
Faculty of Science and Engineering
Manchester Metropolitan University
Manchester
UK

Rashid O. Kadara
Faculty of Science and Engineering
Manchester Metropolitan University
Manchester
UK

ISSN 2191-530X ISSN 2191-5318 (electronic)
SpringerBriefs in Applied Sciences and Technology
ISBN 978-3-319-25191-2 ISBN 978-3-319-25193-6 (eBook)
DOI 10.1007/978-3-319-25193-6

Library of Congress Control Number: 2015950919

Springer Cham Heidelberg New York Dordrecht London

Printed on acid-free paper

Springer International Publishing AG Switzerland is part of Springer Science+Business Media (www.springer.com)

Contents

About this Book

Screen-printing is routinely utilised to fabricate a range of useful electrochemical architectures and is also used in a plethora of areas in both industry and academia. Screen-printing allows the design of next generation electrochemical sensing platforms and allows the translation of proven laboratory-based approaches to be scaled up and commercially applied. In order to undertake this, there are no readily available text book or literature reviews that will actually allow one to set about the task of screen-printing and its technique, and implementation is closely guarded within the field and is seen to the outsider as a “black art”. To those skilled in the art, screen-printing allows the mass production of novel and useful electrochemical architectures that due to the fabrication processes are cost effective yet highly reproducible and yield significant electrical benefits. The numbers of academics and industrialists using screen-printing technology are vast and global pursuit.

This springer brief will be of interest to both academics and industrialists embarking for the first time on screen-printing from those wanting to successfully design, fabricate and implement electrochemical-based architectures and eventually mass produce these, to those who have a basic understanding of the process and want to advance their technical knowledge and skills in this area.

Chapter 1
Introduction and Current Applications of Screen-Printed Electrochemical Architectures

Abstract This chapter provides a general introduction to screen-printed electrochemical platforms with emphasis on pertinent advantageous electrode designs reported within the academic literature.

1.1 History of Screen-Printed Electrodes

Within early electrochemical experiments the utilisation of solid metallic electrodes was a necessity, however over time attention has changed towards the reduction of manufacturing costs, as electrochemistry comes to the forefront of technology. Over time a focus shifted towards the use of carbon based materials within electrochemistry, with the introduction of cheap and affordable approaches such as carbon paste electrodes, which reduce cost significantly but can lack reproducibility [1]. Since the early 1990s the fabrication of electrode circuits via different printing approaches such as pad-printing, roll-to-roll and screen-printing have been utilised for the use within electrochemistry. Each of these printing methods offers inherent advantages and disadvantages. For example, pad-printing offers a thin-film transfer that can be used in an electrochemical setup; however this process is not ideal for mass production of electrode systems and therefore has been regularly deemed a pre-requisite to the screen-printing technology. The screen-printing process has the ability for the mass production of highly reproducible electrode setups [2]. It is with consideration of this that screen-printed electrodes have revolutionised the field due to their capability to bridge the gap between laboratory experiments with in-field implementation [3, 4]. This is of course compounded by the billion dollar (per annum) glucose sensing market which has benefited from the incorporation of screen-printed electrodes since it now allows individuals to be able to measure their blood glucose levels at home, where a result is instantly realised; without recourse

C.W. Foster et al., *Screen-Printing Electrochemical Architectures*,
SpringerBriefs in Applied Sciences and Technology,
DOI 10.1007/978-3-319-25193-6_1

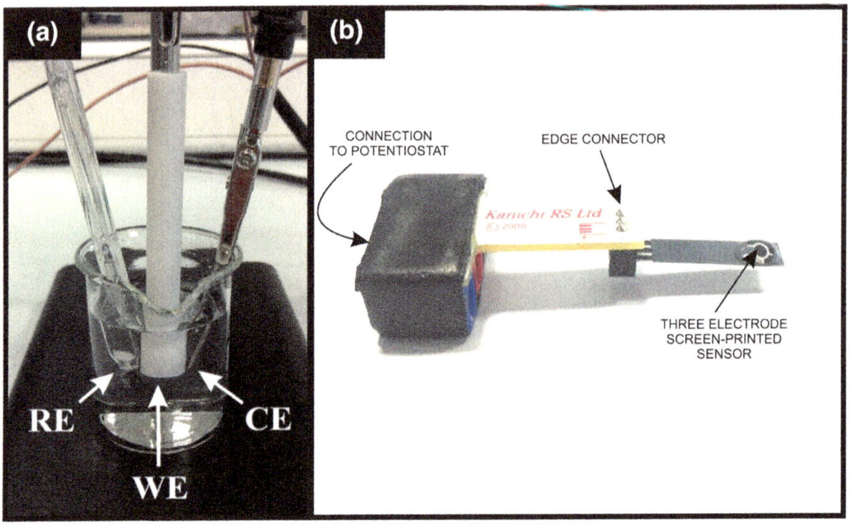

Fig. 1.1 **a** A typical three electrode system utilising an edge pyrolytic plane working electrode, saturated calomel reference electrode and a platinum counter electrode. **b** A screen-printed electrode with a carbon composite working and counter electrode and an Ag/AgCl reference electrode within an electrolyte solution

to visit a hospital/clinic [5, 6]. This technological approach permits the mass production of highly reproducible electrode configurations that possess an excellent scales of economy [6, 7]. Such electrode designs offer improvements in sensitivity, signal-to-noise ratios and reduced sample volumes, producing potential replacements for conventional (solid and re-usable) electrode substrates. Further to this, the ease of the mass production of screen-printed sensors enables their use as one-shot sensors, allowing possible contamination to be avoided, and alleviate the need for electrode pre-treatment as is the case for solid electrodes prior to their use [3, 8, 9]. Figure 1.1 depicts a traditionally used laboratory based three electrode system compared to a system which has been printed using conductive inks; such a comparison indicates the ability to create electrochemical setups that are portable, cheap and reproducible.

Along with the commercially available electrochemical glucose biosensor, these screen-printed electrodes have been regularly utilised for other bio-sensing applications such as the determination of codeine and diazepam within urine samples and beverages respectively [10, 11], lactate sensing and cholesterol to name a few [12, 13]. Additionally DNA sensing using screen-printed electrode systems has become a large focus over the recent decade with much research focusing upon the quantification of mutagenic DNA bases [14–16]. Since the developed sensors are low cost, mass produced, and disposable this allows for the rapid non-intrusive detection of radiation damaged DNA [14]. Other interesting biological applications include the determination of cancer biomarkers, with focus upon an array of proteins

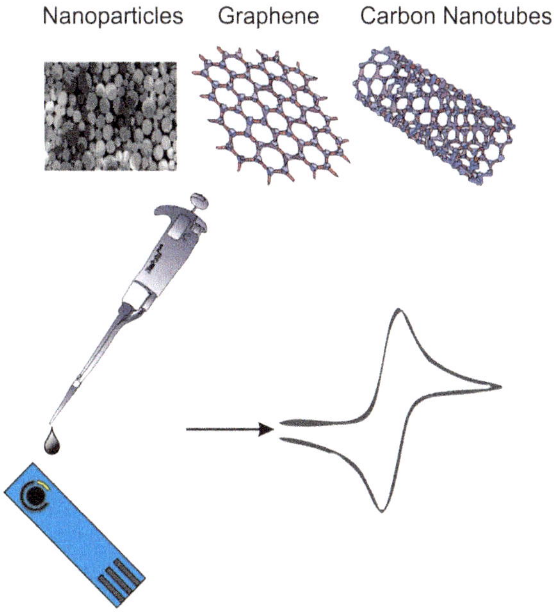

Fig. 1.2 Example of a typical drop-cast fabrication upon screen-printed systems and the corresponding electrochemical measurement of a target analyte. Reproduced from Ref. [18]

released pre-tumour within the body [17]. Figure 1.2 is a typical drop-cast method of modifying the surface of screen-printed systems with a range of modifiers. In this case a range of useful materials can additionally be immobilised upon screen-printed electrodes. In these cases biological elements (such as horseradish peroxidase and graphene oxide etc.) can be incorporated within the chosen modifier [18].

It is noted that this method of fabrication is the most common and simplistic technique of modifying a screen-printed electrode surface. However this approach can potentially offer a low reproducibility and often the films created are not in fact films [19]. Utilising the printing technology the production of reproducible and consistent electrode surfaces are acquired. The ability to create a printing media containing such electrode modifiers gives the user the freedom to produce an even coverage upon the surface [8, 20–22]. Shown in Fig. 1.3 is a typical screen-printed sensor with a printed multi-walled nanotube layer, as the modification is controlled by the parameters dictated within the printing process the creation of a reproducible response for the detection of dopamine is presented [15, 23].

Not only can these systems be applied to biological systems but the incorporation as analytical sensors for an array of electrochemically active toxins, additives and drugs to name a few, has been explored. It is common that these screen-printed electrodes do not have the same limit of detection as other extremely expensive equipment; however the possibility for the creation of point-of-care sensors is an exciting and rapidly evolving area of research [24–26]. For example, regular detection of 'legal-highs' has been continuously successful as the electroactive compounds give intense and quantitative signals, for a potential light-weight handheld sensor [27]. Furthermore these sensors have also been applied within

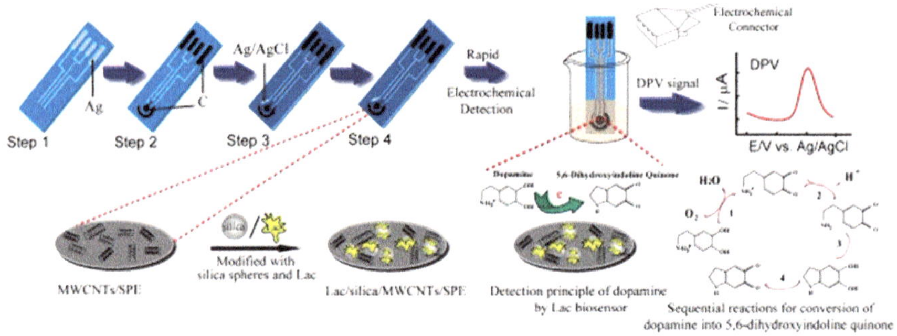

Fig. 1.3 Schematic illustration of the screen-printed configuration, the procedures used in the process of MWCNTs-doped SPE fabrication. Reproduced from Ref. [19]

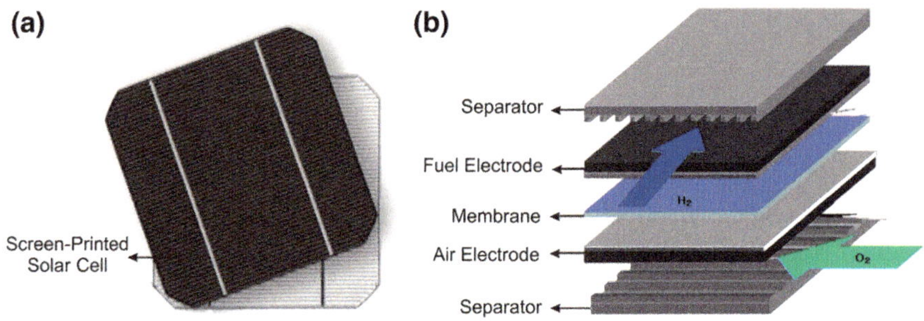

Fig. 1.4 Optical image (**a**) and schematic diagram (**b**) of a screen-printed solar cell and fuel cell respectively. Adapted from Ref. [34, 35]

electrophoretic miniaturised devices due to their benefits such as significant improvement in analysis times, lower consumption of reagents and samples, flexibility, and procedural simplicity [28].

In addition these screen-printed systems can be regularly applied as potential solar cells (Fig. 1.4a), with the incorporation of photocatalytic surface groups, in the form of nanoparticles. It is clear within the literature that screen-printing can directly print onto these substrates easily and can create an even surface coverage [29–31]. Remaining within the area of the renewable energy, focus upon the utilisation of screen-printed electrodes within a potential fuel cell (Fig. 1.4b) has also been studied by simply screen-printing a minimal amount of the cathode and anode material, potentially reducing the overall fabrication and servicing costs of the *future* fuel cell technology [32, 33].

Along with the need for the creation of energy, the storage of the proposed energy must be upheld and improved for the continuation of technological advancement. In this situation, these novel and exciting electrode systems have also been applied towards the exciting capacitative capabilities of the conductive inks [36, 37].

Hallam et al. [38] utilised screen-printing as a new methodology to fabricate flexible and reproducible supercapacitors via the elimination of the interface between the charge collectors and working electrodes, ultimately lead to a higher power density obtained at the printed systems.

With such array of applications the literature is compounded with new additions of research utilising screen-printed electrodes; interested readers are directed to the following authoritative reviews [3, 39, 40].

1.2 Screen-Printed Designs

Additionally, from the tailoring of the working electrode material through the use of screen-printing techniques, the electrode geometry is also readily manipulated through the incorporation of screen-printing technologies for the fabrication of new and exciting electrochemical sensing platforms. When designing such platforms it is convenient to have electrodes which do not require a large sample size, therefore in general the required three-electrodes are fabricated as close as possible and also reduce resistance which can potentially distort the measured voltammetry; the most common electrode configuration is the 3 mm working electrode with on-board reference and counter (see Fig. 1.1).

In addition to macroelectrode architectures, microelectrodes can be designed and fabricate to mimic that of true microelectrode behaviour (i.e. near elimination of contributions from planar diffusion rather than solely radial/convergent diffusion) and further enhancements in the electroanalytical sensing performances. The conversion from macro to microelectrode systems within electrochemical applications, possess some significant advantages, such as: smaller diffusion layer, greater mass transport, lower transient current and more resistive to the effects the ohmic drop. These systems also have benefits within the fabrication method, due to the minimal ink utilised during screen-print.

A variant on the traditional co-planar disc-shaped microelectrode which continues to gather further interest within current literature is the microband electrode. Band electrodes are fabricated to be macroscopic in-length but microscopic in width. These electrode configurations have been reported to offer the additional advantage of allowing larger currents to be obtained compared to a microdisc due to the increased electrode area, whilst the width of the band is still maintained in the micrometre range to ensure convergent diffusion is still dominant. Generally these have been fabricated via the screen-print method however the utilisation of gold and platinum inks are generally to printed upon a ceramic substrate, offering less user-ability and applicability within-the-field. Upon this printed electrode, a Pyrex glass slide is used to define a thin microband system [41, 42]. In adaptation to this Craston et al. [43] reported the use of a screen-printed conductive ink (Au/Pt) with an additional polymer layer upon a ceramic layer, this is then etched in a perpendicular direction of the print (shown in Fig. 1.5), creating a microband-*like* electrode system. Recently an adapted approach within electroanalytical applications has been carried

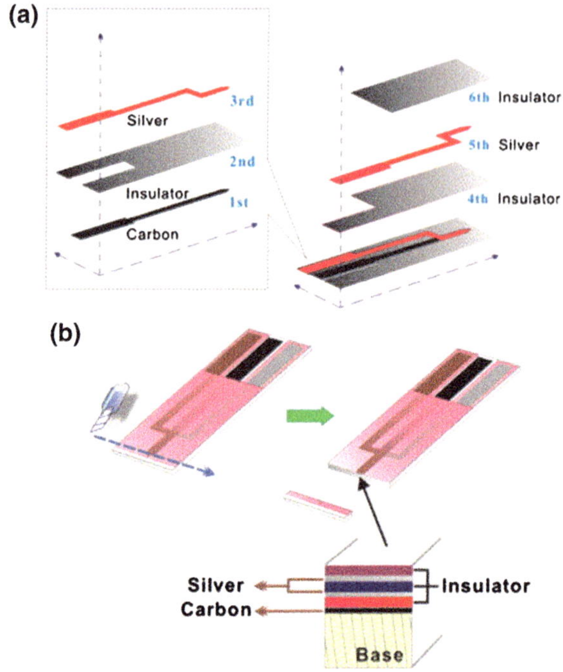

Fig. 1.5 **a** A schematic of a layer-by-layer structure of the screen-printed ultra-microelectrode assembly. **b** Cross-sectional diagram the electrode system with a built-in three-electrode configuration. Reproduced with permission from Ref. [48]

out by Hart and Zen independently [44, 45] where ultramicrobands are fabricated in the aforementioned manner and were targeted towards the sensing of lead in acetate leachates [45] and determination of glutathione [46]. However, the reproducibility of the electrodes fabricated in this manner is questionable [47].

Nonetheless upon utilisation of the appropriate printing parameters this can be completed reproducibly and accurately. Metters et al. [47] reported the use of a V-Mesh (Vecry Mesh) (explained further within Sect. 2.1) for a defined and reproducible screen-printed 50 microband electrode, such electrode was employed as a electroanalytical sensor [47]. In addition to this, Souza et al. have also screen-printed a 100 micron band (shown in Fig. 1.6) electrode fabricated using a polyester screen, with high reproducibility within electroanalytical studies [21, 26].

In consideration of the above mentioned limitations of macroelectrodes, the creation of screen-printed array electrodes can be realised for the potential multiple detection of intriguing analytes. Over recent decades these microelectrode arrays can significantly increase the sensitivity of electroanalytical methodologies, due to the increase number of microelectrode disks present within the design [49].

Rusling et al. [50] have created an immunoassay that can detect a range of cancer biomarker proteins upon a simple microarray electrode design. Shown in Fig. 1.7 is an example of such screen-printed electrode microelectrode array and how they can be modified with antibodies which can potentially bind to the cancer biomarker proteins [50–52].

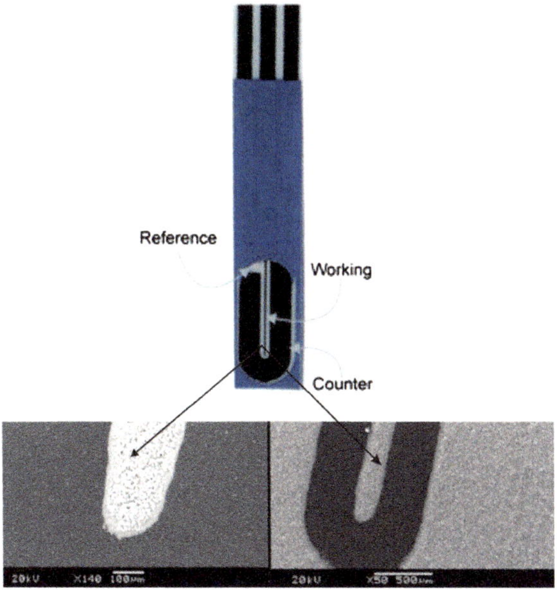

Fig. 1.6 Optical and SEM imaging of a 100 μm screen-printed band electrode

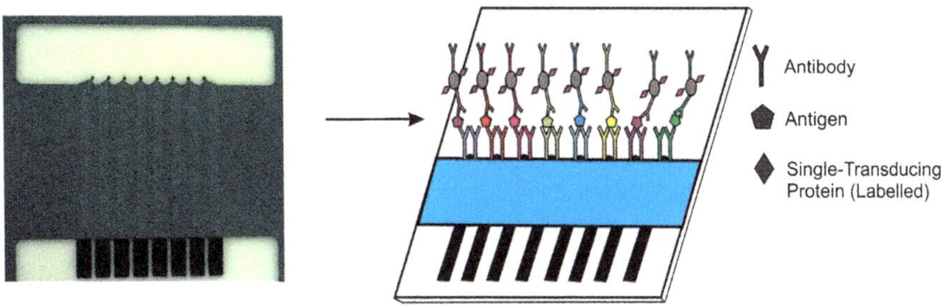

Fig. 1.7 Optical image of a microarray screen-printed electrode and schematic image of an immunoassay

These arrays allow for the multiple detection of analytes; however, utilisation of recessed arrays can also be accomplished via the screen-printing process. Recessed microelectrodes and arrays can be fabricated via screen-printing a thicker dielectric layer around the microelectrode, allowing for an extended distance between each electrode and thus create more diffusional independent voltammetric responses (i.e. sigmoidal responses), creating advantages such as greater sensitivity, increased current density and improved signal-to-noise ratio [53].

1.3 Printing Media

The utilisation of a printing medium (i.e. ink or paste) within the screen-printing process is vital for the transfer of the printed design, typically these inks or pastes are fabricated using conductive particles within a solvent/binder mixture to allow the transfer of the particulate matter onto the chosen substrate. Such inks can be made readily available and are extremely beneficial to the user as they possess the ability to be manipulated to provide different working electrode compositions.

Wang et al. [54] indicates that the choice of carbon ink should rely upon the application at hand, due to the differences within the electrochemical signals seen with a range of inks [54]. Such other examples of printable inks, include the fabrication of platinum [55] and gold [56] screen-printed sensors which have been applied towards the electroanalytical sensing of chromium species (VI and II) in the case of the gold sensor, and both hydrazine and hydrogen peroxide for the case of the platinum sensor [55, 56]. Crucially, it was determined that the requirement for electrode potential cycling prior to utilisation (as is the case for bulk noble metal macro electrodes in order to form an oxide upon the electrode surface) was alleviated in the case of the screen-printed sensors owing to the noble metal utilised within the screen-printing process being in the form of an oxide [35, 36]. Clearly, the removal of such a preparatory step offers significant benefits when considering the development of sensors intended for use outside of the laboratory environment where rapid and facile analysis is imperative (one-step analysis). However mentioned previously the metallic options can be costly and the need for carbon based materials has become necessary. In response to this, the utilisation of graphitic based inks for an assortment of applications has been realised. Not only do these inks replicate the electrochemistry of typically used electrodes such as glassy carbon (GC) and edge plane pyrolytic graphite (EPPG) but they are manufactured at a fraction of the cost. In addition an array of carbon materials can be used, such as carbon-nanotubes [20], graphene [8], mediated carbon structures [57] and nanoparticles [58] to name a few.

Having the ability to alter the electron transfer capabilities at an electrode surface has a beneficial effect upon the analytical applications. Shown within Fig. 1.8 compares voltammetric behaviour of unmodified screen-printed electrodes with that of polymeric modified screen-printed electrode; upon the metallic plating of these systems the electrochemistry will be dominated by the metallic properties rather than the underlying electrode.

Choudhry et al. have shown beneficial modification of various electro-catalytic metals where the bespoke modified electrode surface performs as a micron-sized template [59]. When the metallic ensemble is used over that of an electrolytically modified macroelectrode, a dramatic change within the mass transport occurs enabling a low limit of detection and potential improvements in the analytical performance even with a minimal amount of the electrode surface being covered by the ensemble.

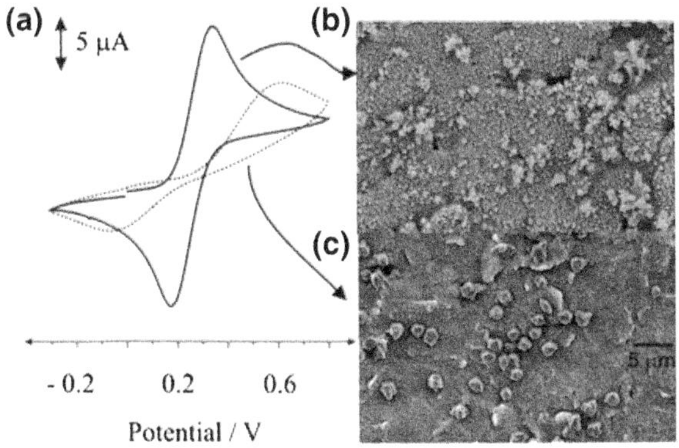

Fig. 1.8 Cyclic voltammetric profiles (**a**) obtained in 1 mM potassium ferrocyanide in 1 M KCl using the standard electrochemical platform (*solid line*) with that of a bespoke electrochemical platform (*dashed line*). Scans recorded at 100 mV s^{-1} versus SCE. SEM images of the copper plated standard (**b**) and bespoke (**c**) screen printed electrochemical platforms. Note the change in the ink used in the electrode surface gives rise to different morphologies. Reproduced from Ref. [59] with permission from Elsevier

In addition to pure metallic modifications as reported above, the utilisation of metallic oxides can be used. Bismuth oxide is a common oxide form that can be placed within a printable ink for advantageous detection of metal ions, with such modification being *akin* to that of highly toxic mercury electrode systems. Unlike that of an in situ or ex situ modification the fabrication process can create extremely reproducible areas of catalytic material within the composite, rather than on the surface. Such electrode systems mimic that of the bismuth film electrodes and therefore can be utilised for the array of applications of these electrode systems [3].

1.4 Substrate

One aspect of screen-printing technology, particularly in the case of the fabrication of electrochemical devices which is often overlooked, is the selection of the electrode substrate on which the ink is printed upon. Screen-printed electrodes are generally printed upon ceramic [60, 61] or plastic substrates [62–64] and the need for ultra-flexible sensors has arisen due to the possibility of using the screen-printed electrodes not only just within the laboratory environment but within the scientific field outside its confines applied "into-the-field". Previous elegant work has seen "biosensors in briefs" being used to sense chemicals such as hydrogen peroxide and NADH [65]; a case in which ultra-flexible sensors are fundamental in achieving an ideal electrode for such applications. Wang et al. [66] developed wearable electrochemical sensors on underwater garments comprised of the synthetic rubber neoprene. The neoprene-based sensor was evaluated towards the voltammetric

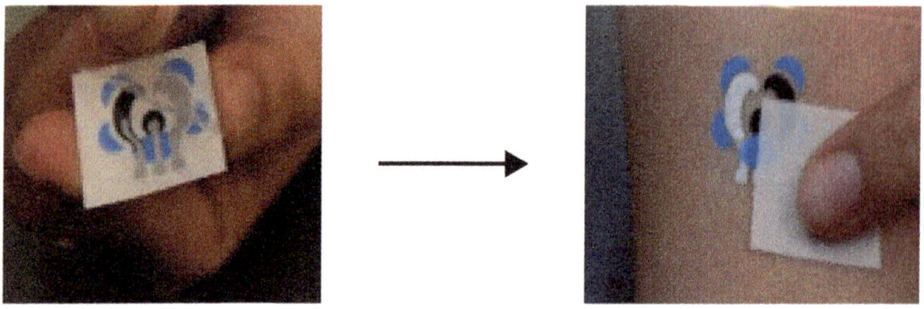

Fig. 1.9 Optical images of the electrochemical tattoo upon the substrate and skin. Reproduced from Ref. [67]

detection of trace heavy metal contaminants and nitro-aromatic explosives in seawater samples, with further applications involving the first example of enzyme (tyrosinase) immobilization on a wearable substrate towards the amperometric biosensing of phenolic contaminants in seawater also being described. Furthermore, depicted in Fig. 1.9 is the screen-printed fabrication of electrochemical tattoos. Wang and co-workers have demonstrated that temporary transfer tattoo (T3) electrochemical sensors can be fabricated for physiological and security monitoring of chemical constituents leading to the demonstration of 'electronic skin', within such report they look at the effect of stretching and friction created on the skin [67]. Additionally studies by Wang and co-workers have studied the analysis of sodium within perspiration upon the epidermis, utilising a similar setup as shown in Fig. 1.9 [68].

Other work has explored the effect of mechanical contortion and stress on polymeric sensors comprised of Mylar, polyethylene naphthalate and Kapton which was found to be able to withstand such mechanical stress and still function electrochemically [63]. Recently we have studied the different electrochemical properties of ultra-flexible screen-printed paper-based sensors (standard desktop printing paper) against that of the more traditional polyester-based screen-printed electrodes determining that little deviation with regard to the electrochemical capabilities of the sensors is noted between the two variants, even after extreme mechanically contortion [69, 70]. Generally a good substrate is simply classed as it having a good adhesion of the screen-printing ink; the current diversity of substrates, as shown above, is fascinating.

References

1. Ravichandran K, Baldwin RP (1981) J Electroanal Chem Interfacial Electrochem 126:293–300
2. Mooring L, Karousos NG, Livingstone C, Davis J, Wildgoose GG, Wilkins SJ, Compton RG (2005) Sens Actuat B: Chem 107:491–496
3. Metters JP, Kadara RO, Banks CE (2011) Analyst 136:1067–1076
4. Honeychurch KC, Hart JP (2003) Trends Anal Chem 22:456–469

5. Heller A, Feldman B (2008) Chem Rev 108:2482–2505
6. Schlesinger M (2013) Mod Aspects Electrochem: Appl Electrochem Med 56:83–84
7. Hayat A, Marty JL (2014) Sensors 14:10432–10453
8. Randviir EP, Brownson DAC, Metters JP, Kadara RO, Banks CE (2014) Phys Chem Chem Phys 16:4598–4611
9. Kolliopoulos AV, Metters JP, Banks CE (2013) Anal Methods 5:3490–3496
10. Honeychurch KC, Crew A, Northall H, Radbourne S, Davies O, Newman S, Hart JP (2013) Talanta 116:300–307
11. Shih Y, Zen JM, Yang HH (2002) J Pharm Biomed Anal 29:827–833
12. Wang Y, Xu H, Zhang J, Li G (2008) Sensors (Basel, Switzerland) 8:2043–2081
13. Carrara S, Shumyantseva VV, Archakov AI, Samorì B (2008) Biosens Bioelectron 24:148–150
14. Wang J, Rivas G, Ozsoz M, Grant DH, Cai X, Parrado C (1997) Anal Chem 69:1457–1460
15. Ren R, Leng C, Zhang S (2010) Biosens Bioelectron 25:2089–2094
16. Liu J, Su B, Lagger G, Tacchini P, Girault HH (2006) Anal Chem 78:6879–6884
17. Taleat Z, Khoshroo A, Mazloum-Ardakani M (2014) Microchim Acta 181:865–891
18. Couto RAS, Lima JLFC, Quinaz MB (2015) Talanta. doi:10.1016/j.talanta.2015.06.011
19. Li Y, Zhang L, Li M, Pan Z, Li D (2012) Chem Cent J 6:103–103
20. Randviir EP, Metters JP, Stainton J, Banks CE (2013) Analyst 138:2970–2981
21. Ruas de Souza AP, Bertotti M, Foster CW, Banks CE (2015) Electroanalysis. doi:10.1002/elan.201500155
22. Foster CW, Metters JP, Kampouris DK, Banks CE (2014) Electroanalysis 26:262–274
23. Wang J, Chen L (1995) Anal Chem 67:3824–3827
24. Amine A, Mohammadi H, Bourais I, Palleschi G (2006) Biosens Bioelectron 21:1405–1423
25. Nie Z, Nijhuis CA, Gong J, Chen X, Kumachev A, Martinez AW, Narovlyansky M, Whitesides GM (2010) Lab Chip 10:477–483
26. Ruas de Souza AP, Foster CW, Kolliopoulos AV, Bertotti M, Banks CE (2015) Analyst 140:4130–4136
27. Smith JP, Metters JP, Irving C, Sutcliffe OB, Banks CE (2014) Analyst 139:389–400
28. Randviir EP, Banks CE (2015) Electrophoresis. doi:10.1002/elps.201500153
29. Jalali M, Siavash Moakhar R, Kushwaha A, Goh GKL, Riahi-Noori N, Sadrnezhaad SK (2015) Electrochim Acta 169:395–401
30. Ito S, Chen P, Comte P, Nazeeruddin MK, Liska P, Péchy P, Grätzel M (2007) Prog Photovoltaics Res Appl 15:603–612
31. Yurkiv V, Constantin G, Hornes A, Gondolini A, Mercadelli E, Sanson A, Dessemond L, Costa R (2015) J Power Sour 287:58–67
32. Ge X, Huang X, Zhang Y, Lu Z, Xu J, Chen K, Dong D, Liu Z, Miao J, Su W (2006) J Power Sour 159:1048–1050
33. Kluczowski R, Krauz M, Kawalec M, Ouweltjes JP (2014) J Power Sour 268:752–757
34. Micro-co Ltd. http://www.e-microtec.co.jp/topics/ev_pvexpo09E.html. Accessed 08 Sept 2015
35. Bullis K (2015) Efficient cheap solar cells. http://www.technologyreview.com/news/410862/efficient-cheap-solar-cells/. Accessed 08 Sept 2015
36. Xu Y, Schwab MG, Strudwick AJ, Hennig I, Feng X, Wu Z, Müllen K (2013) Adv Energy Mater 3:1035–1040
37. Ji X, Hallam PM, Houssein SM, Kadara R, Lang L, Banks CE (2012) RSC Adv 2:1508–1515
38. Hallam PM, Gomez-Mingot M, Kampouris DK, Banks CE (2012) RSC Adv 2:6672–6679
39. Thiyagarajan N, Chang J-L, Senthilkumar K, Zen J-M (2014) Electrochem Commun 38:86–90
40. Hart JP, Wring SA (1997) TrAC Trends Anal Chem 16:89–103
41. Chang J-L, Zen J-M (2006) Electroanalysis 18:941–946
42. Authier L, Grossiord C, Brossier P, Limoges B (2001) Anal Chem 73:4450–4456
43. Craston DH, Jones CP, Williams DE, El Murr N (1991) Talanta 38:17–26
44. Prasad KS, Chen JC, Ay C, Zen JM (2007) Sens Actuat B: Chem 123:715–719
45. Honeychurch KC, Al-Berezanchi S, Hart JP (2011) Talanta 84:717–723
46. Honeychurch KC, Hart JP (2012) Adv Anal Chem 2:46–52

47. Metters JP, Kadara RO, Banks CE (2013) Analyst 138:2516–2521
48. Chang J-L, Zen J-M (2006) Electrochem Commun 8:571–576
49. Davies TJ, Ward-Jones S, Banks CE, del Campo J, Mas R, Muñoz FX, Compton RG (2005) J Electroanal Chem 585:51–62
50. Rusling JF, Bishop GW, Doan N, Papadimitrakopoulos F (2014) J Mater Chem B Mater Biol Med 2. doi:10.1039/C1033TB21323D
51. Otieno BA, Krause CE, Latus A, Chikkaveeraiah BV, Faria RC, Rusling JF (2014) Biosens Bioelectron 53:268–274
52. Rusling JF, Kumar CV, Gutkind JS, Patel V (2010) The Analyst 135:2496–2511
53. Kadara RO, Jenkinson N, Banks CE (2009) Sens Actuat B: Chem 142:342–346
54. Wang J, Tian B, Nascimento VB, Angnes L (1998) Electrochim Acta 43:3459–3465
55. Metters JP, Tan F, Kadara RO, Banks CE (2012) Anal Methods 4:1272–1277
56. Metters JP, Kadara RO, Banks CE (2012) Analyst 137:896–902
57. Foster CW, Pillay J, Metters JP, Banks CE (2014) Sensors 14:21905–21922
58. Tiwari I, Singh M, Gupta M, Metters JP, Banks CE (2015) Anal Methods 7:2020–2027
59. Choudhry NA, Kadara RO, Jenkinson N, Banks CE (2010) Electrochem Commun 12:406–409
60. Salazara P, Martín M, O'Neill RD, Roche R, González-Mora JL (2012) Electroanal Chem 674:48–56
61. Matemadombo F, Apetrei C, Nyokong T, Rodríguez-Méndez ML, Antonio de Saja J (2012) Sens Actuat B: Chem 166:457–466
62. Ramdani O, Metters JP, Figueiredo-Filho LCS, Fatibello-Filho O, Banks CE (2013) Analyst 138:1053–1059
63. Cai J, Cizak K, Long B, McAferty K, Campbell CG, Allee DR, Vogt BD, La Belle J, Wang J (2009) Sens Actuat B: Chem 137:379–385
64. Foster CW, Ruas de Souza AP, Metters JP, Bertotti M, Banks CE (2015) Analyst. doi:10.1039/C5AN01692D
65. Yang Y, Chuang M, Lou S, Wang J (2010) Analyst 135:1230–1234
66. Malzahn K, Windmiller JR, Valdes-Ramirez G, Shoning MJ, Wang J (2011) Analyst 136:2912–2917
67. Windmiller JR, Bandodkar AJ, Valdes-Ramirez G, Parkhomovsky S, Martinez AG, Wang J (2012) Chem Commun 48:6794–6796
68. Bandodkar AJ, Molinnus D, Mirza O, Guinovart T, Windmiller JR, Valdés-Ramírez G, Andrade FJ, Schöning MJ, Wang J (2014) Biosens Bioelectron 54:603–609
69. Foster CW, Metters JP, Banks CE (2013) Electroanalysis 25:2275–2282
70. Metters JP, Houssein SM, Kampouris DK, Banks CE (2013) Anal Methods 5:103–110

Chapter 2
Fundamentals of Screen-Printing Electrochemical Architectures

Abstract This chapter introduces the background theory on the screen-printing process.

2.1 Screen-Printing Process

The technique of screen-printing is described as the production of thick film hybrids, for applications within many areas of the scientific community, such as circuit boards and electrochemical systems to name a few. Due to the nature of screen-printing the creation of mass-produced thick films can be realised, with the utilisation of relatively cheap and simple designs this process possesses excellent scales of economy. Additionally it is noted that due to simplicity of the machinery used throughout, this process can be altered and changed upon a fundamental understanding of screen-printing.

The process of screen-printing typically consists of five prerequisites to ensure identically reproducible thick films, these are as follows:

- Suitable printing medium
- Mesh screen with an embedded stencil design
- Substrate to print upon
- Flexible and resilient squeegee
- Secure base to prevent movement of substrate within the process.

It is important to note that suitable printing machinery is desired to ensure the desired level of reproducibility is achieved with a high throughput. Nevertheless there are situations where the printing machinery can impede and become inauspicious to the overall screen-print, it is with these considerations that it is solely reliant on the methodology.

C.W. Foster et al., *Screen-Printing Electrochemical Architectures*,
SpringerBriefs in Applied Sciences and Technology,
DOI 10.1007/978-3-319-25193-6_2

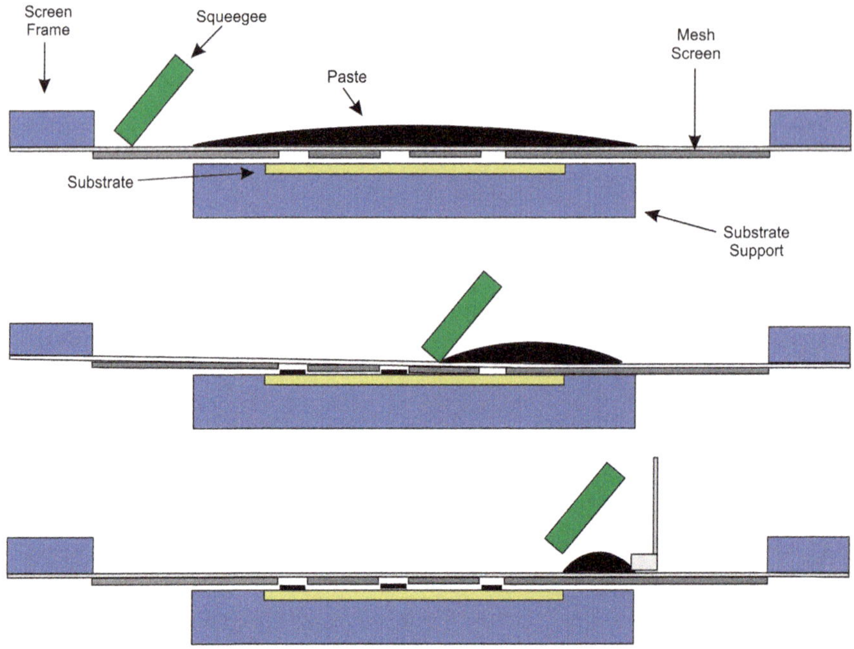

Fig. 2.1 Schematic of the printing process

Upon integration of the above mentioned pre-requisites, a print cycle can occur; Fig. 2.1 represents the process in three simple steps. The first step consists of placement of the printing media (i.e. ink, emulsion or paste) upon the mesh screen, where it is visible that the screen is not in contact with the substrate. Such contact could potentially damage or spoil the print as the substrate would uncontrollably "snap-off" the screen, creating an unclean print. The contact only occurs when the squeegee applies pressure over the screen forcing the printing medium through the stencil design, creating the desired pattern or design in a controlled and efficient manner.

2.2 Selection of Screen-Printing Equipment

As mentioned previously the prerequisites are vital for the screen-printing process, it is additionally important the selection of this equipment is crucial for maximising reproducibility and outputs. It is with this section that consideration of the screen, squeegee, flood/distribution bar and printing media are discussed.

2.2.1 The Screen

When utilising and designing the screen there are three factors to consider such as the screen frame, stencil design and the screen-mesh. The screen frame typically is fabricated utilising wood or metal, however due to their similarity in pricing they tend to be made from metal, creating a robust and safe foundation. The material used must be able to withstand the pressure created from the tension of the mesh, which in some cases can exceed values of 80 kg. The frame size can be varied however in all cases the screens tend to be either rectangular or square depending on the size of the machine and design at hand. It is due to this that there is no specific ratio or value that corresponds to the perfect design, however it does refer to the inside dimensions of the screen (as the frame is a support for the open mesh area of the screen, calculated by Eq. 2.1). There are mainly two types of frames for the screen-printing process these are lightweight cast aluminium and extruded aluminium, dependent on the machine of choice the appropriate composition can be selected.

$$Open\ Area = \frac{(mesh\ area)^2}{(wire\ diameter + mesh\ opening)^2} \times 100\,\% \qquad (2.1)$$

The screen-mesh has many interesting properties that allow for a perfect screen-printed design, as it acts as a support network to hold the stencil-design in place at all times during the print cycle, even after the pressure has been applied by the squeegee. These characteristics originate from its design and manufacture, the choice of an appropriate material for the mesh can be vital for a precise and detailed screen-print. Shown in Fig. 2.2 is an optical image of a screen-mesh showing the woven network of the mesh. In this screen the electrodes will be fabricated; note the distance in the size of the mesh, to allow a certain amount of ink to transfer onto the substrate.

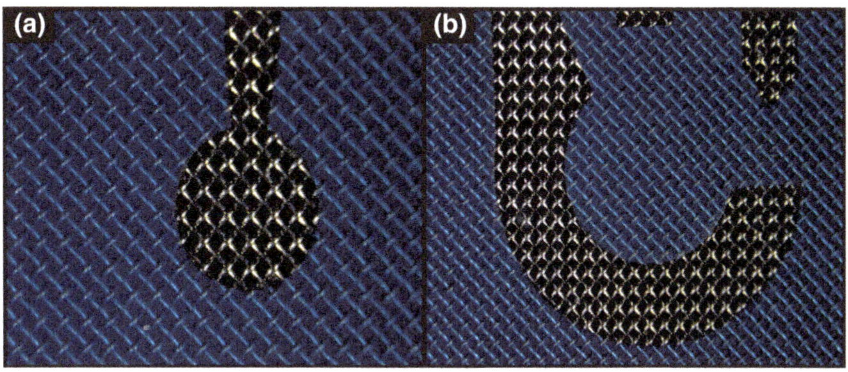

Fig. 2.2 Optical image of a stainless steel screen-mesh, for a working electrode (*left image*) and counter electrode (*right image*)

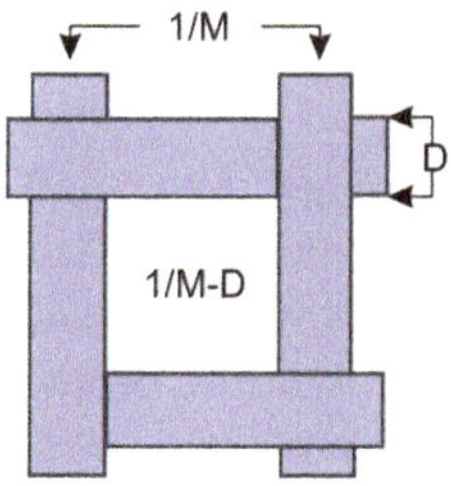

Fig. 2.3 Schematic of a mesh opening

A factor that must be considered upon creation of the mesh screen is the mesh count, M, i.e. the amount of wires per unit of length as shown in Fig. 2.3.

In conjunction with size of the mesh count and wire diameter, D, the mesh opening can be calculated via Eq. 2.2. The mesh opening is a vital measurement that influences the amount and size of the ink that can be passed through the screen.

$$O = \frac{1}{M} - D \tag{2.2}$$

Many manufacturers prefer to use three types of materials these being nylon, polyester and stainless steel; represented in Table 2.1 are the benefits of utilising each material.

Along with careful consideration of Table 2.1, other vital observations must be carried out. For example, the minimum line width that can be printed must be three times the mesh thread diameter (i.e. large mesh threads cannot print small designs). In addition the mesh diameter must also be three times larger than the particulate size of the printing medium, to allow suitable passage of the media. Furthermore to the above mentioned mesh designs, it is possible to utilise a V-mesh, which consists of a mesh created from Vecry which is a sheathed filament surrounding a liquid crystal based polymer core. This mesh provides the required strength for a screen, the combination of thin, flexible fibres and the weaving process result in flat

Table 2.1 Representation of the capabilities of the three materials for the design of mesh screens

Screen requirements	Polyester	Stainless steel	Nylon
Flexibility	1	2	3
Resilience	1	2	3
Percentage of open area	2	1	2
Stability of print size	2	1	3
Damage from squeegee	2	3	1
Accidental damage	2	3	1
Cost	1	3	1
Minimal snap-off from large areas	2	1	3

Ranked 1–3 which corresponds to the worst (3) and best (1) in its category

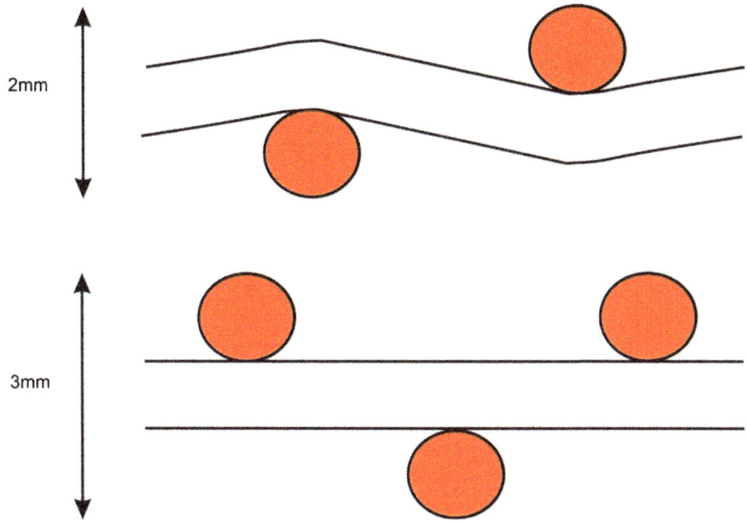

Fig. 2.4 Representation of the effect on heat rolling of polyester and nylon mesh screens

filament intersections. Additionally, the smoother thread surfaces further enhance paste release. Such screens can exhibit higher levels of tension allowing a less-snap off, thus reducing the risk of faulty screen-prints. The thickness of the screen is additionally vital for the printing process, in most cases the overall screen-mesh thickness will be double that of the wire diameter. Typically within the fabrication process of the polyester and nylon mesh screens the mesh is woven and then passed through heat rollers which weld the wires together creating a bound matrix of mesh wires. In the case of stainless steel meshes the metallic properties of the steel allows for instant mesh strength upon weaving of the wires, therefore this additional heating step is no longer required. Represented within Fig. 2.4 are the resulting wire mesh thicknesses upon utilisation of the two fabrication methods. It is noticeable that a distortion of the wire mesh occurs upon utilisation of the heat rollers, therefore reducing the wire diameter from 3 to 2 mm. It is important to note that the stainless steel mesh thicknesses can vary an extraordinary amount due to the nature of the stiffness and hardness of the steel wires.

As mentioned previously the amount of printing medium that can be transferred relies on the opening in the screen and the thickness, the volume of paste (V) can be calculated using Eq. 2.3. Assuming the screen thickness is double that of the wire diameter:

$$V = (2D)\left(\frac{1}{M} - D\right)^2 \tag{2.3}$$

In order for a successful print with a minimal snap-off the screen-mesh must be held at a specific tension which must be sufficient to allow the screen to peel away from the substrate in a controlled and reproducible fashion. Nonetheless excessive tension upon the mesh can cause detrimental and costly damage. It is with manipulation of these tensile qualities that a perfect mesh can be created. It is duly noted that screens that are highly taught will give the best control of the snap-off and therefore reproducibility of the print. On the other hand, highly taught screens are more likely to be susceptible to unwanted or accidental damage due to their excessive tension. Therefore when utilising stainless steel screens, elongation of 0.9 % will allow the screen to revert back to its original state, with an additional 0.1 % as leverage. It is possible for a mesh screen to give a perfectly defined and reproducible print at values of half that of the elongation limit, considering that an alteration to the print-gap between the screen and substrate is incurred, for the perfect snap-off.

2.2.2 *The Squeegee*

During a screen-print cycle a highly important factor is the constant movement of the printing media towards the substrate, at a sufficient speed and pressure, this is accomplished by the utilisation of a squeegee. As described in Sect. 2.1 the squeegee applies pressure to forcing the ink through the stencil design, creating an even print upon the substrate in a controlled manner.

For the squeegee to complete its necessary tasks, it must be fabricated with materials that are flexible and resilient. The most common material used throughout industry is Polyurethane, it is general practise that squeegees possess an extremely long life and on average can achieve prints of 20,000 before any visible damage occurs. However utilising stainless steel meshes will decrease this as the surface friction applied during cycle will eventually damage the point of contact. During the fabrication of the squeegees many grades of softness are created, it is generally accepted that the softest material creates a larger contact with the screen; therefore this is the more efficient grade to use. Depicted in Fig. 2.5 is the potential effect of squeegee damage towards the coverage of printing media during the print cycle, it is clear that upon the snap-off from the substrate the print is incomplete and irregular. Additionally the squeegee must be at least 10 mm wider than that of the print area, it is apparent that a larger squeegee will create a smaller the natural print gap, as there is more pressure applied on the screen-mesh potentially damaging the screen irrevocably (shown in Fig. 2.6).

The angle at which the squeegee applies its pressure is also a vital condition for the successful transfer of print media through the stencil design. If the angle is too high the transfer of the printing medium will not be sufficient to fill the print area, therefore creating a thinner application of ink/paste to the substrate below. Additionally if the squeegee is at a much shallower angle the hydrodynamic pressure increases, therefore transferring too much of the ink/paste onto the substrate and potentially blocking and hindering the controlled snap-off. Figure 2.7 exhibits a selection of angles and their advantages/disadvantages, to the screen-printing process.

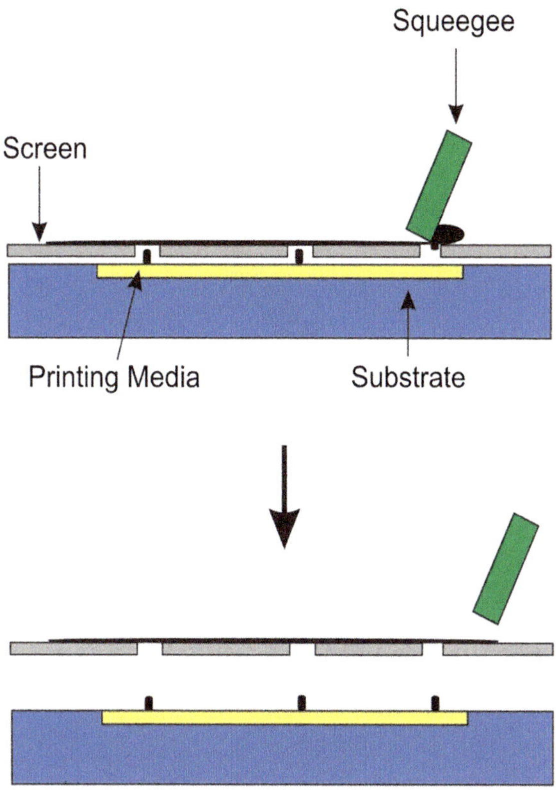

Fig. 2.5 Schematics of the effect of a damaged squeegee during the print cycle, upon the transfer of ink to the substrate; not that the ink has been pushed through the screen correctly and the deposited ink does not reflect the original defined image

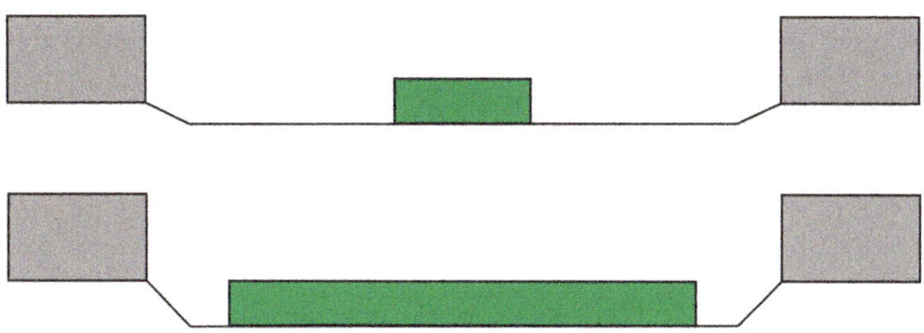

Fig. 2.6 Representation of the squeegee size and its effect upon the screen-mesh, where it can be seen that upon utilisation of a larger/heavier squeegee a naturally smaller print gap is created

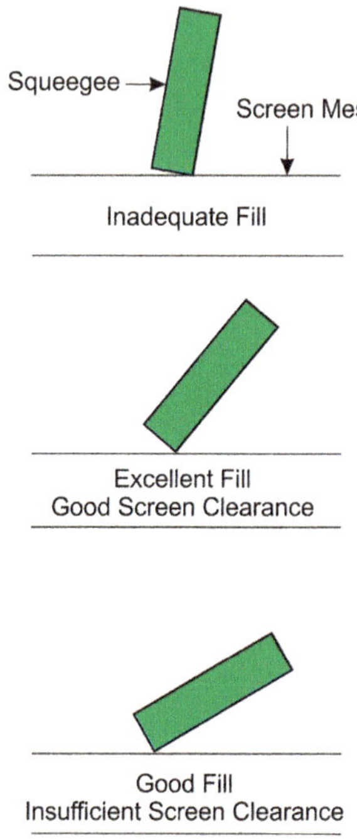

Fig. 2.7 Angles of the squeegee and their effect upon the screen-print

2.2.3 *The Flood Bar/Distributor*

The distribution of ink/paste is another vital process of the print cycle for a successful print. Typically flood bars are designed from stainless steel and will be slightly longer than the squeegee being utilised, it is then fixed behind the squeegee and upon the first transition of the print cycle the flood bar will transport the ink/paste over the print area. It is vital that the flood bar is slightly above or touching the screen-mesh, so that it is constantly in touching distance of the ink. In many printing situations the squeegee will remove the excess ink/paste back to the starting position, and the process can be repeated.

2.2.4 *Printing Medium*

The screen-printed process allows the user to fabricate a variety of geometric designs and shapes; such ability requires the utilisation of a durable, compatible printing medium. Terminology of such media can range from a dye to a simple ink

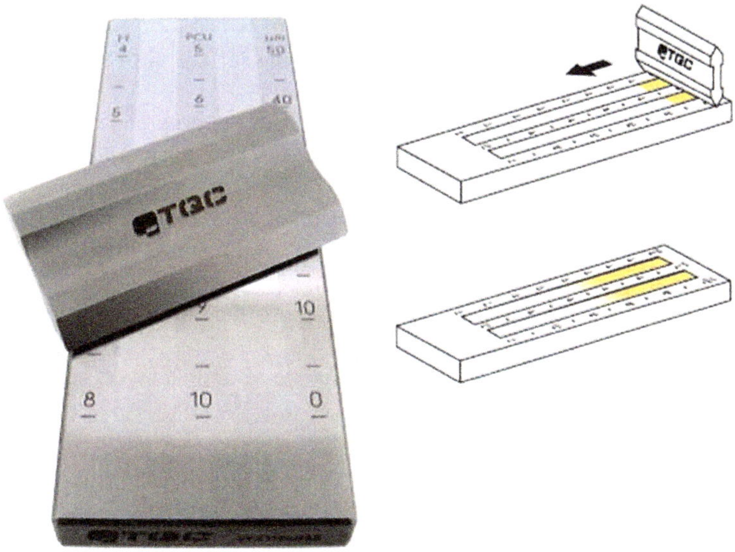

Fig. 2.8 Optical images of fineness of grind gauge. Reproduced from Ref. [1]

or paste, but in most cases they all will possess the same composition. The viscosity of these 'inks' will determine how successful the print is, as due to the nature of screen-printing the ink must be passed through a specific shape (stencil) keeping its geometric design. The formulation of the specialist inks tend to consist of two components: suitable pigments for the application at hand and an appropriate amount of solvent/binder ratio creating the perfect transport of the pigment.

Upon application of screen-printing some designs or screen-prints may require viscosities that are higher than the ideal, as the majority of quality control issues arise when the ink is more viscous, therefore it is important to ensure that the operation procedure should be modified via optimising each step of the manufacture. The ideal screen-printing ink should require no forcing into the open area of the screen, flow readily when moved by the flood bar and not dry within the screen-mesh during the operation. To achieve this perfect consistency, prior mixing of the ink must occur until a smooth fluidic composition is achieved.

In general examination of the composition is performed by utilising a fineness of grind (FOG) gauge (shown in Fig. 2.8) for the determination of dispersion, particle size, and fineness of many materials like inks, lacquers, pigments, filler, chocolate to name a few. In this situation, the materials being tested are inks. The gauge may also be used to indicate the presence of undesired large particles in these materials. Such tool has an attached scrapper, which pulls the material along the sloping groove machined onto the top surface of the gauge. The value for fineness of grind is obtained directly from a scale engraved into the gauge.

Standard Operating Procedures for the Fineness of Grind Gauge

1. Place the gauge on a flat, horizontal and non-slip surface, with the zero mark on the scale closet to the user.
2. Place a considerable amount of material (ink) in the deep end of each groove.
3. Place the scrapper on the surface of the gauge behind the material (ink), which is at the deepest groove.
4. Use both hands to hold the scrapper and pull along the length of the gauge at a constant speed apply sufficient pressure to clean the excess material (ink) from the edge of the gauge.
5. Stop at a point beyond the zero depth and assess the drawn out material within the next 3 s.
6. Note: This avoids inaccurate testing due to evaporation of solvents from materials.
7. The material (ink) should be viewed at right angles to the length of the groove and at an angle of 20–30 with the surface of the gauge.
8. Find the first position across the groove 3 mm wide which contains 3–10 particles/streaks/scratches of material (ink).
9. Read the position on the scale and record this value.
10. Perform the test three times; afterwards calculate the average value of the result. The average value is the fineness of grind of the material (ink).

The FOG test can be used as a QA/QC measurement for inks used in the screen-printing process.

2.2.5 *Curing of the Ink*

The nature of the printing media requires a curing step within the fabrication of the design. Generally many failures in the fabrication of conductive inks are reliant upon the poor selection of solvents. It is with this in mind that consideration of the solvents utilised within the 'ink' allow for an ideal curing time and temperature to envisage a situation depicted within Fig. 2.9; where it is clear that upon the curing step of the ink the volume starts to decrease, leaving behind a fully conductive 'stack' upon the substrate where the polymer in the ink formulation holds the structure of the electrode surface.

In many cases the manufacturer of the ink will provide the solvents that are the best ratio for working and curing properties, therefore if an amendment is requested consideration of the curing procedure must be endured. Note that an ink has a range of solvents varying from slow and fast evaporation time, to ensure a controlled

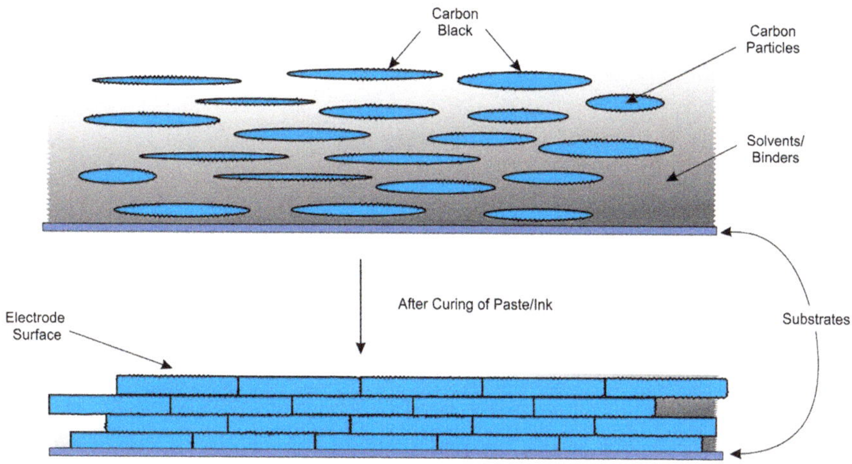

Fig. 2.9 Representation of curing of the printing medium upon the conductive layers

drying process resulting in a reproducible electrode surface. It is important to note that the solvent is duly there to create a fluidic support for the conductive paste and thus in most cases a solvent with a lower boiling point would be ideal.

Reference

1. TQC USA. https://www.tqc-usa.com/en/products/article/grindometers-fineness-of-grind-gauges. Accessed 08 Nov 2015

Chapter 3
Fabricating Screen-Printed Electrochemical Architectures: Successful Design and Fabrication

Abstract This chapter introduces a practical guide to how to screen-print.

3.1 A Practical Guide to Screen-Printing

3.1.1 Standard Operating Procedure for Screen-Printing

In consultation with the prior section upon choosing the appropriate screen mesh, design, squeegee and printing medium, the fabrication of the screen-printed design can begin. Within this section a precise standard operating procedure for the manufacture of screen-printed electrodes is now described.

(1) It is important to note that screen-printing is a completely clean process, therefore full protective clothing must be utilised. Therefore a full body suit, plastic gloves and hairnet are a necessity, typically in a clean room environment.

C.W. Foster et al., *Screen-Printing Electrochemical Architectures*,
SpringerBriefs in Applied Sciences and Technology,
DOI 10.1007/978-3-319-25193-6_3

(2) The first step is to choose the suitable screen-design for the desired electrode system. It is essential to remember that screen-printed electrodes require three-layers of different ink, therefore *ENSURE* the appropriate layer is selected.

(3) Pick up the screen holding the frame (not the mesh). As so not to damage or contaminate it.

(4) To guarantee the successful transfer of ink to the substrate the tension of the screen must be measured, to determine if the perfect snap-off will occur. (Note: the value for this depends on the starting tension measurement and screen mesh specification.) This parameter can also be used to inform you when a new screen is required.

(5) Ensuring that the screen is the right way round (with the top of the screen facing downwards), carefully place and secure the screen into the printing machine.

(6) After securing the screen into the machine, the printing medium can now be prepared. In general, due to the viscosity of the printing paste/ink extensive mixing of the ink must occur to create a fluidic homogenous mixture. This can be either done vigorously by hand or mechanically.

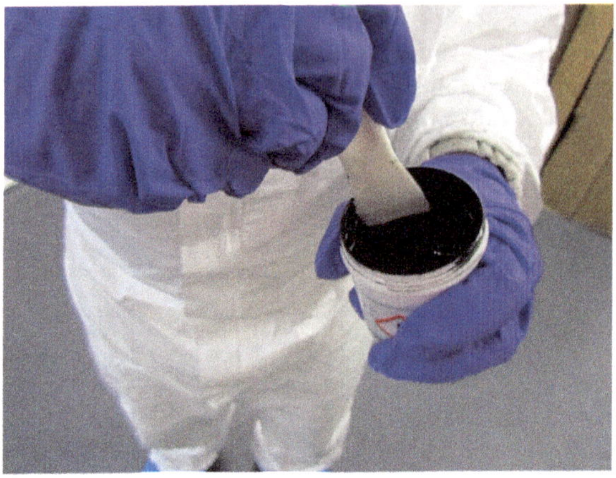

(7) Once the viscosity of the paste is reduced, it is now ready for application upon the screen within the machine. It is vital that the ink is applied next to the screen design, with a suitable amount to print the desired quantity of electrodes, thus reducing the potential waste and costs.

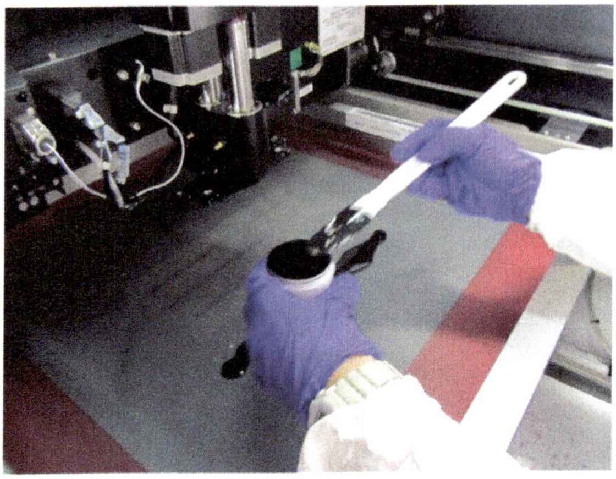

(8) Now that the printing medium has been applied, the placement of the squeegee and flood bar can proceed. Firstly, place the flood bar on the supporting printer head, ensuring that the bolts are securely fastened. Secondly, place the clean squeegee in front of the flood bar and secure the bolts in a similar fashion to the first step. It is important to note that to stop contamination from different inks/pastes the utilisation of different floor bars and squeegees is considered best practice for each ink used.

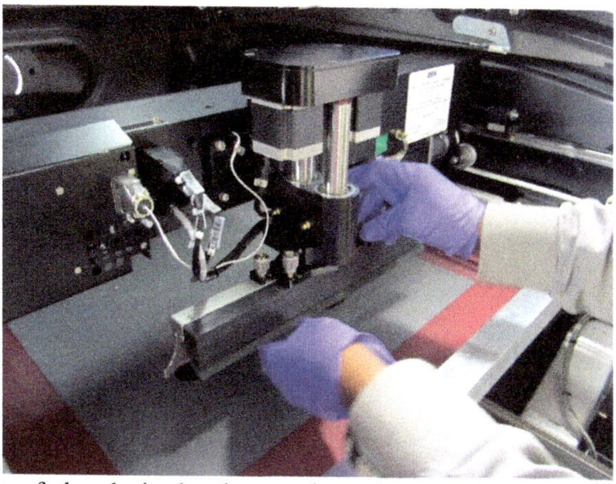

(9) Placement of the desired substrate into the machine can now occur. It is extremely necessary that you are as reproducible as possible, so that each layer overlays successfully.

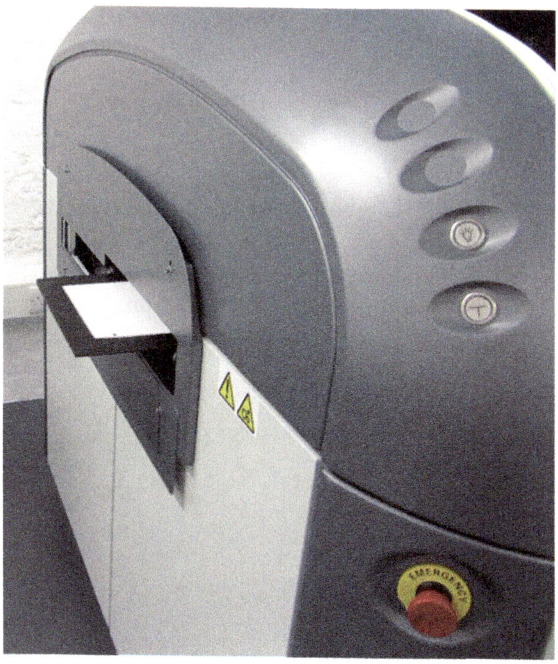

(10) Now the screen-print cycle can occur, select run and click the 'GO' button, upon this instruction the machine will pull the substrate into the machine and proceed with the cycle. Firstly, the flood bar lowers and then drags the ink over the mesh design, the squeegee will then return to the starting position applying a pressure which forces the ink to pass through the stencil design.

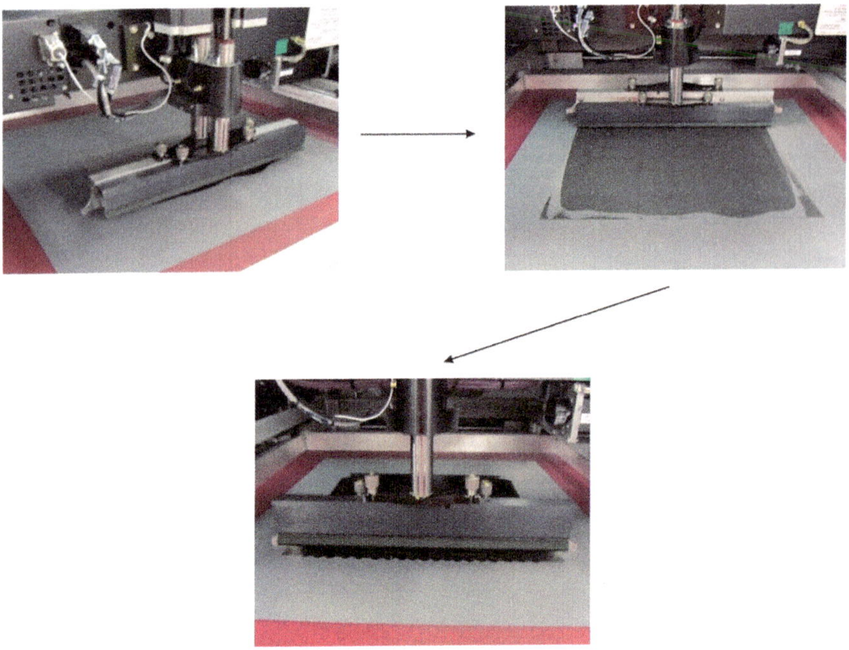

(11) When the cycle has finished the substrate will be released by the machine, with the stencil design transferred on the top of it. A thorough inspection should indicate any unprinted areas or regions of uneven coverages. If the print is compromised consult the troubleshooting section.

(12) Repeat until the required amount of electrodes has been accomplished. To complete the process, curing of the ink must occur, within an oven at a step temperature and time; usually given by the ink manufacture.

(13) To increase productivity it is good practice to clean the equipment used throughout the screen-printing process, while the electrodes cure. First remove the squeegee and the flood bar, being careful not to damage the screen. After remove the screen (handling it in the same fashion as previously mentioned), and place within a well ventilated area and clean using an appropriate solvent. When cleaning the screen you may apply slight force on the underside as this does not affect the stencil design, however you must take serious care when cleaning the top of the screen, which contains the stencil design.

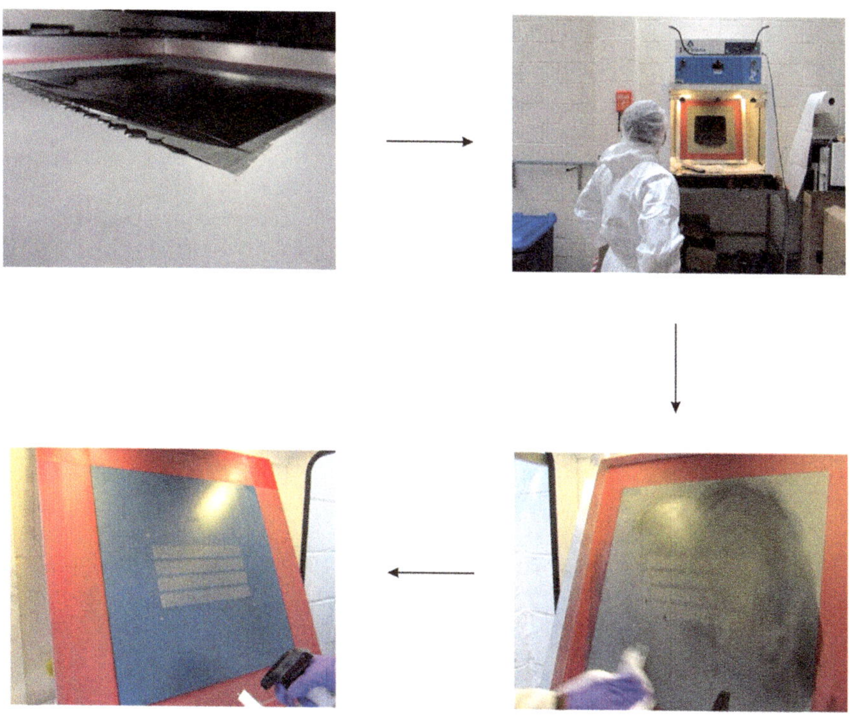

Upon completion of the first layer, movement onto the second, third and in some cases fourth can begin, utilising the required screens the layers can be produced in a reproducible and quick fashion.

3.2 Screen-Printing Troubleshooting Guide

Presented below is a detailed troubleshooting table indicating the problem, the cause and how to resolve many commons issues when utilising the screen-printing process (Table 3.1).

Table 3.1 Troubleshooting guide for the screen-printing process

Issue	Equipment causing the problem	Cause of the problem	How to resolve issue
Screen-print incomplete	Squeegee	Not parallel with the screen	Turn blade round or renew blade
		Gap is too large between the squeegee and the screen	Decrease the gap between the squeegee and screen
		The squeegee is not wide enough for full coverage	Minimum squeegee must have a width that is 10 mm larger on each side of screen image
	Printing medium	Viscosity of printing medium too thick for successfully filling of screen-design	Use a recommended screen; with a less viscous ink or emulsion
		Dried ink within the screen-design	Remove the screen and clean with appropriate solvent
	Screen	Incorrect positioning of the print area	Correct so that the print design is in the middle of substrate
		Print gap too large or small	Change to the appropriate screen-substrate gap
		Screen mesh too fine	Change to screen recommended for work
Erroneous and undesirable screen-print	Squeegee	Areas of un-printed ink, could be caused by dusty/dirty components	Raise printhead and clean the underside removing any dirty or dust
		Image may contain streaks check squeegee edge for worn/damaged behaviour	Fit new squeegee blade
		Printing medium not cleared from the screen mesh due to the squeegee pressure being too low	Increase pressure in small steps until a clear track is obtained and then increase slightly

(continued)

Table 3.1 (continued)

Issue	Equipment causing the problem	Cause of the problem	How to resolve issue
		Print smudging from the squeegee to screen gap being too small	Increase the gap until the screen peels away from substrate in a controlled manner
		Squeegee pressure too heavy causing poor definition of the edge of the design	Reduce squeegee pressure in small steps until good print is obtained, then increase pressure slightly
	Printing medium	Stringing and serrating of the image edges due to the ink being too thin	Use thicker or more viscous inks
		Thin deposit of ink upon the print	Use less viscous ink
	Screen	Mesh damaged and affecting the print area	Fit a new screen
		Print image serrated at edge, and has poor definition as the screen-mesh is too coarse or the screen open area is too small	Change to screen to one with a recommended finer mesh/use recommended screen with thicker emulsion
		Deposit too thin, from the screen mesh being too fine	Use a coarser mesh
Substrate issues	Substrate	Contaminated printing surfaces or excessively bowed	Clean substrates thoroughly before printing. Increase vacuum, mechanical gripping, reduce squeegee pressure or reject substrate
		Vacuum is insufficient to hold substrates when very viscous pastes are used	Check whether: (i) Vacuum pipe not fully secured to unions (ii) Holes at registered printing position are blocked (iii) Filter of pump is blocked (iv) Pump exhaust pipe is restricted (v) Other leaks exist
		Print gap too small	Set print gap correctly

Chapter 4
Quality Control/Quality Assurance Analysis of Electrochemical Screen-Printed Sensors

Abstract Now that you have successfully fabricated your electrochemical architectures you will need to characterise and test their electrochemical performance. Consequently, this chapter provides a brief overview to electrochemistry introducing the necessary theory that is required to perform experiments that will allow one to electrochemically characterise and test the fabricated sensors.

4.1 Cyclic Voltammetry

This technique is the most widely utilised for obtaining quantitative information about electrochemical reactions. This approach involves applying a potential over a set timeframe. The term *voltammetry* derives from volt-am(père)-metry and one should not confuse this with voltametry (note change in spelling!)—written with a single "*m*"—which is controlled-current potentiometric titration as defined by IUPAC. It tenders the rapid identification of *redox* potentials distinctive to the electroactive species, providing considerable information about the thermodynamics of a redox process, kinetics of heterogeneous electron-transfer reactions, coupled electrochemical reactions or adsorption processes (note that a general introduction to voltammetry is given and other textbooks should be considered for a rigours overview) [1–4]. Cyclic voltammetry consists of scanning (linearly) the potential of the working electrode using a triangular potential wave form (Fig. 4.1).

The potential is swept from E_1 to E_2 and the rate at which this is achieved is the voltammetric scan rate (or the gradient of the line), as shown in Fig. 4.1 (V/s). In this case, if the potential is stopped at E_2, this is known as a linear sweep experiment. If the scan is returned back to E_1, a full potential cycle, this is known as cyclic voltammetry. Depending on the information sought, either single or multiple cycles can be performed. For the duration of the potential sweep, the potentiostat measures the resulting current that arises via the applied voltage (potential). The plot of current versus potential (voltage) is termed a 'cyclic voltammogram'. A cyclic voltammogram is complex and dependent on time along with many other

C.W. Foster et al., *Screen-Printing Electrochemical Architectures*,
SpringerBriefs in Applied Sciences and Technology,
DOI 10.1007/978-3-319-25193-6_4

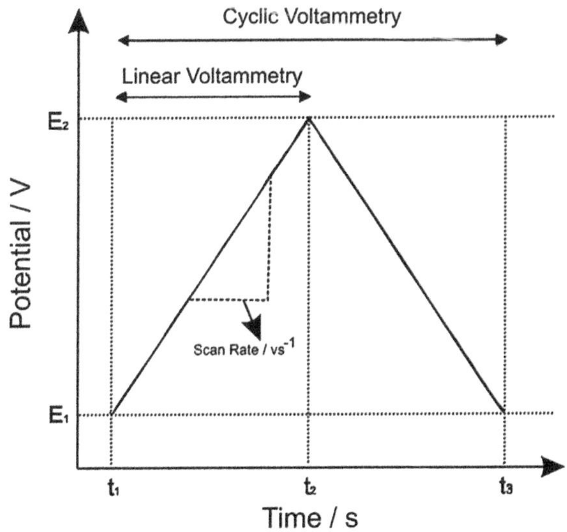

Fig. 4.1 Potential—time profiles used to perform linear sweep and cyclic voltammetry

physical and chemical properties. Figure 4.2 shows a typical cyclic voltammetric curve (or CV) for the case of the electrochemical process where a voltammetric potential is applied and the current monitored which gives rise to the unique profile presented in Fig. 4.2. Note that characteristics of the voltammogram which are routinely monitored and reported are the peak height (I_P) and the potential at which the peak occurs (E_P).

The magnitude of the cyclic voltammogram and E_P can be utilised for quality control purposes, to verify that a screen-printed batch conforms to an expected output; those outside this range can be discarded.

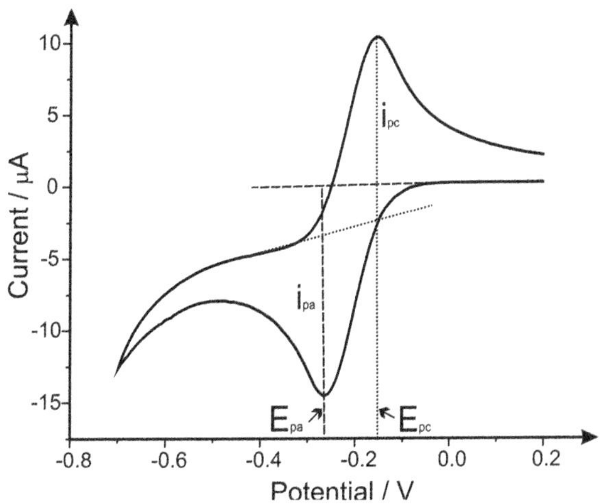

Fig. 4.2 Typical cyclic voltammogram depicting the peak position E_P and peak height I_P

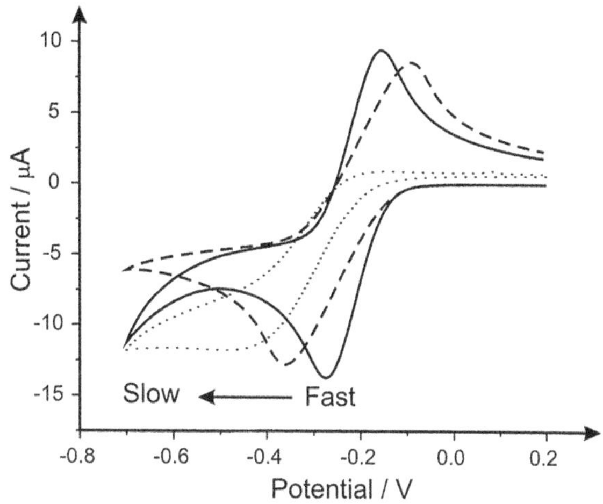

Fig. 4.3 Cyclic voltammograms for reversible (*solid line*), quasi-reversible (*dashed line*) and irreversible (*dotted line*) electron transfer

Shown in Fig. 4.3 is the case of different heterogeneous electron transfer rates, that is, *reversible*, *quasi-reversible* and *irreversible*, each giving rise to unique voltammetric profiles. The physical processes responsible for the characteristic shape of a 'reversible' voltammogram, for the process, $A + ne^- \longrightarrow B$ is based on (i) Fick's laws and (ii) Nernst's laws:

$$\frac{\partial[A]}{\partial t} = D\frac{\partial^2[A]}{\partial x^2}; \quad \frac{[A]_0}{[B]_0} = e^{\frac{nF\eta}{RT}} \tag{4.1}$$

where the Nernst law is written in the exponential form. If we consider the case of the reversible cyclic voltammogram, (shown in Fig. 4.4–solid line) at positive potentials no current flows, as the potential is driven negatively, the electrochemical rate constant (k^0) becomes large enough and the current rises as the potential becomes increasingly negative. Last, at the more negative potential the current passes a maxima and decreases. Note that in cyclic voltammetry the diffusion layer continually expands. Readers are directed to the following text books for a rigorous analysis of cyclic voltammetry [1–4].

In the 'reversible' limit the electrode kinetics are so 'fast' (relative to the rate of mass transport—see later) that Nernstian equilibrium is attained at the electrode surface throughout the voltammogram with concentrations of A and B at the electrode surface governed by the Nernst equation:

$$E = E^0_{red}(A/B) + \frac{RT}{F} In\frac{[B]_0}{[A]_0} \tag{4.2}$$

where E is now the applied potential which defines the ratio of the surface concentrations $[A]_0$ and $[B]_0$ once $E^0_{red}(A/B)$ is specified, R is the gas constant, F is the Faraday constant.

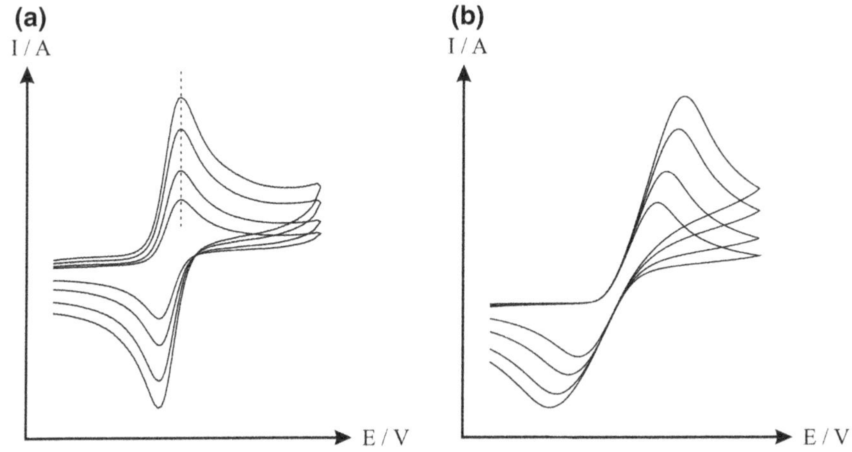

Fig. 4.4 Reversible (**a**) and irreversible (**b**) cyclic voltammetric responses. Note the shift of the peak maxima with scan rate

In the case of an electrochemically reversible process *with* fast electron transfer, the peak-to-peak separation $\Delta E_p = (E_p^{ox} - E_p^{red})$ is relatively small at the reversible limit, where $\Delta E_p = 2.218RT/nF$, corresponding to a value of *ca.* 57 mV (at 298 K where $n = 1$). For the case of n electrons, the wave-shape of the voltammogram can be characterised by:

$$E_p - E_{1/2} = 2.218\frac{RT}{nF} \tag{4.3}$$

where $E_{1/2}$ correspond to potential at which half the peak current is observed and n is the number of electrons presented within the electrochemical reaction.

The magnitude of the voltammetric current (I_p^{Rev}) observed at a macroelectrode is governed by the following Randles–Ševćik equation, for a fully reversible case:

$$I_p^{\text{Rev}} = \pm 0.446nFAC(nFD\upsilon/RT)^{1/2} \tag{4.4}$$

where the $\pm$ sign is used to indicate an oxidation or reductive process respectively though the equation is usually devoid of such sign, υ is the scan rate applied and D is the diffusion coefficient of the analyte present within the solution. The voltammetric diagnosis that the electrochemical process is undergoing a reversible heterogeneous charge transfer process is given by Eq. (4.3) where ΔE_p is independent of the applied voltammetric scan rate and: $I_p^{ox}/I_p^{red} = 1$.

The question is, how can you determine if your observed voltammetry corresponds to this range? A key diagnostic is a scan rate study. As shown in Eq. (4.4), the peak height (I_P) is proportional to the applied voltammetric scan rate and a plot of I_p^{Rev} against $\upsilon^{1/2}$ should be linear. Figure 4.4 depicts typical voltammetric profiles resulting from applying a range of scan rates. It is evident that each voltammetric signature is the same but that the current increases with increasing scan rate as predicted by Eq. (4.4). It is important to note that when the position of the current maximum occurs at the same potential; this peak maximum, which does *not* shift in potential with scan rate, is characteristic of electrode reactions which exhibit rapid electron transfer kinetics, usually termed reversible electron transfer reactions.

The formal potential can be found as the mid-way between the two voltammetric peaks comprising the voltammogram:

$$E_f^0 = \left(E_p^{ox} + E_p^{red}\right)/2 \tag{4.5}$$

assuming that the diffusion coefficients of the reactant and product are equal.

Also shown in Fig. 4.4 is the cyclic voltammetric response for an irreversible electrochemical couple (where the ΔE_P is larger than that observed for the reversible and quasi-reversible case) where appreciable over-potentials are required to drive the reaction, as evidenced by the peak height occurring at a greater potential than that seen for the reversible case.

In Fig. 4.4 it is evident that as the standard electrochemical rate constant, k^o, is either fast or slow, termed 'electrochemically reversible' or 'electrochemically irreversible' respectively, changes in the observed voltammetry are striking. It is important to note that these are relative terms and that they are in relation to the rate of mass transport to the electrode surface. The mass transport coefficient, m_T, is given by:

$$m_T = \sqrt{D/(RT/F\upsilon)} \tag{4.6}$$

The distinction between fast and slow electrode kinetics relates to the prevailing rate of mass transport given by '$k^o \gg m_T$' indicating electrochemical reversibility or '$k^o \ll m_T$' indicating electrochemical irreversibility. Matsuda and Ayabe [5] introduce the parameter, ζ, given by:

$$\zeta = k^o/(FD\upsilon/RT)^{1/2} \tag{4.7}$$

where the following ranges are identified at a stationary macroelectrode: '$\zeta \geq 15$' corresponds to the reversible limit, '$15 > \zeta > 10^{-3}$' corresponds to the quasi-reversible limit and '$\zeta \leq 10^{-3}$' corresponds to the irreversible limit. Thus returning to Fig. 4.4, we have three cases, reversible, quasi-reversible and irreversible which are all related to the rate of mass transport. In reversible reactions the electron transfer rate is, at all potentials, greater than the rate of mass transport and the peak potential is independent of the applied voltammetric scan rate (shown in Fig. 4.4a). In the case of quasi-reversible the rate of electron transfer becomes comparable to the mass transport rate. In this regime, the peak potentials increase with the applied scan rate. Last, it is obvious that for the irreversible case the electron transfer rates are smaller than the rate of mass transport; the summary by Matsuda and Ayabe is extremely useful [5].

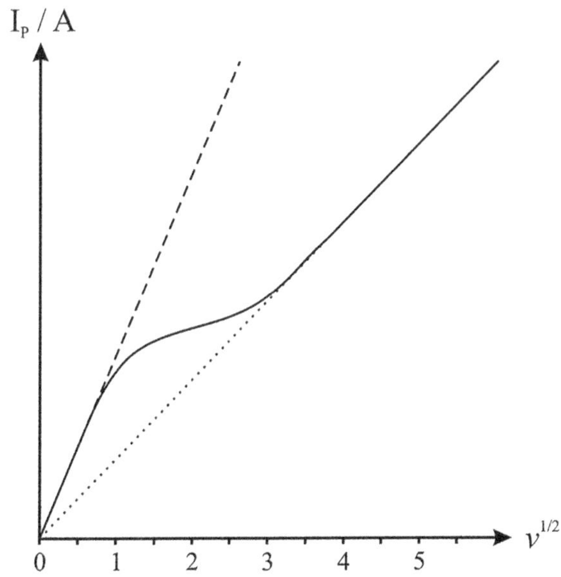

Fig. 4.5 Transition from a reversible to an irreversible process with increasing scan rate (*solid line*). The *dashed line* indicates a reversible process, while the *dotted line* is that of an irreversible process

The above conditions given by Matsuda and Ayabe show that the observed electrochemical behaviour depends on the applied voltammetric scan rate. In applying various scan rates the diffusion layer thickness dramatically changes, in the case of slow scan rates, the diffusion layer is very thick while at faster scan rates the diffusion layer is relatively thinner. Since the electrochemical process, that is, reversible or irreversible reflects the competition between the electrode kinetics and mass transport, faster scan rates will encourage greater electrochemical irreversibility. This is shown in Fig. 4.5 where upon the application of faster scan rates, there is a clear transition (solid line, Fig. 4.5) from that of reversible towards irreversible behaviour.

At macroelectrodes the Nicholson method is routinely used to estimate the observed standard heterogeneous electron transfer rate for quasi-reversible systems using the following equation [6]:

$$\psi = k^{o}[\pi D n \upsilon F/(RT)]^{-1/2} \tag{4.8}$$

where ψ is the kinetic parameter and is tabulated (see Table 4.1) at a set temperature for a one-step, one electron process as a function of the peak-to-peak separation (ΔE_P) where one determines the variation of ΔE_P with υ and from this, the variation in the ψ. Table 4.1 shows the variation of ΔE_P with ψ for a one-step, one electron process at 298 K and where $\alpha = 0.5$. A plot of ψ against $[\pi D n \upsilon F/(RT)]^{-1/2}$ allows the standard heterogeneous rate transfer constant, k^o to be readily deduced.

Note that there are some restrictions, in that the above method is based on the assumption that electron transfer kinetics are described by the Butler-Volmer formalism, that α is 0.5, the switching potential is 141 mV past the reversible $E_{1/2}$,

Table 4.1 Variation of ΔE_p with ψ at 298 K

ψ	$\Delta E_p \times n$/mV
20	61
7	63
6	64
5	65
4	66
3	68
2	72
1	84
0.75	92
0.50	105
0.35	121
0.25	141
0.10	212

Reproduced with permission from Ref. [6]

and the temperature is 298 K. Lack of strict adherence to most of these factors will lead to only minor errors. However, there is one experimental problem that can be severe: incomplete compensation of solution resistance. As such, measurement error will be low at slow scan rates where currents and *IR* errors are low, generally however, potentiostats help overcome this problem.

Beyond the limits of the Nicholson method, that is where the ΔE_P is > 200 mV (see Table 4.1), a suitable relationship has been reported by Klingler and Kochi [7]:

$$k^o = 2.18[D\alpha n\upsilon F/(RT)]^{1/2} \exp[-(\alpha^2 nF/RT)(E_p^{ox} - E_p^{red})] \tag{4.9}$$

Thus two procedures are available for different ranges of $\Delta E_P \times n$ values, that is for low (Nicholson) and high values (Klingler and Kochi). Lavagnini et al. [8] proposed the following function of $\psi(\Delta E_P)$, which fits Nicholson's data, for practical usage (rather than producing a working curve):

$$\psi = (-0.6288 + 0.0021X)/(1 - 0.017X) \tag{4.10}$$

where $X = \Delta E_P$. For more accurate results in determining k^o, recourse to electrochemical simulation packages is advised. Thus, clearly the k^o, is a useful parameter to benchmark electrodes.

Last, the Randles–Ševćik equation for a quasi-reversible system (at 298 K) is given by:

$$I_P^{quasi} = \pm(2.65 \times 10^5)n^{3/2}ACD^{1/2}\nu^{1/2} \tag{4.11}$$

For an irreversible system (those with slow electron exchange), the individual peaks are reduced in magnitude and widely separated. Figure 4.5 shows a characteristic response where the peak maximum clearly shifts with the applied voltammetric scan rate. Totally irreversible systems are quantitatively characterised by a shift in the peak potential with scan rate as given by:

$$E_{p,c} = E_f^0 - \frac{RT}{\alpha n' F}\left[0.780 + In\frac{D^{1/2}}{k^0} + 0.5In\left(\frac{\alpha n' F \upsilon}{RT}\right)\right] \quad (4.12)$$

where α is the transfer coefficient, n' is the number of electrons transferred per mole before the rate determining step and where E_f^0 is the formal potential. Hence, E_P occurs at potentials higher than E_f^0, with the over potential related to k^o and α (the voltammogram becomes increasingly 'drawn out' as αn decreases). For the case of a fully irreversible electron transfer process, the Randles–Ševićik equation is:

$$I_p^{irrev} = \pm 0.496(\alpha n')^{1/2} nFAC(FD\upsilon/RT)^{1/2} \quad (4.13)$$

where A is the geometric area of the electrode (cm^2), α is the transfer coefficient (usually assumed to be close to 0.5), n is the total number of electrons transferred per molecule in the electrochemical process and n' is the number of electrons transferred per moles before the rate determining step. It is useful to know the generic Randles–Ševićik equation (for stagnant solutions):

$$I_p = -\Upsilon(p)\sqrt{\frac{n^3F^3\upsilon D}{RT}}A[C] \qquad where: \; p = r\sqrt{\frac{nF\upsilon}{RTD}} \quad (4.14)$$

for the case of different electrode geometries:

(1) Planar disc electrode: r = radius, $\Upsilon(p) = 0.446$
(2) Spherical or hemispherical electrode: r = radius, $\Upsilon(p) = 0.446 + 0.752p^{-1}$
(3) For a small disk electrode: r = radius, $\Upsilon(p) = 0.446 + \left(0.840 + 0.433e^{-0.66p} - 0.166e^{-11/p}\right)p^{-1} \sim 0.446 + 4/\pi p^{-1}$
(4) For a cylinder or hemi-cylinder: r = radius, $\Upsilon(p) = 0.446 + 0.344p^{-0.852}$
(5) For a band electrode: $2r$ = width, $\Upsilon(p) = 0.446 + 0.614(1 + 43.6p^2)^{-1} + 1.323p^{0.892} \sim 0.446 + 3.131p^{-0.892}$

The wave-shape for an irreversible reduction is given by: $E_p - E_{1/2} = 1.857\frac{RT}{\alpha F}$, while for an irreversible oxidation it is given by: $E_p - E_{1/2} = 1.857\frac{RT}{(1-\alpha)F}$.

The above analysis allows one to readily determine key parameters (such as E_p, I_p, k^O etc.) with the utilisation of screen-printed electrodes.

4.2 Selection of Redox Probes

In the above case we considered the generic redox example. In order to run the cyclic voltammetric experiment described above one needs to select an appropriate electroactive analyte. Typically these probes are classified into inner- and outer-sphere redox probes. There are a whole range of redox probes that exist and that can be utilised and have been studied on highly ordered pyrolytic graphite and glassy carbon electrodes [9]: $IrCl_6$, $Ru(NH_3)_6$, $Co(phen)_3$, MV, $Fe(phen)_3$, $Fe(CN)_6$, $Co(en)_3$, $Ru(en)_3$, $Fc(COOH)_2$, $Ru(bpy)_3$, $Ru(NH_3)_5py$, Co(sep), $Ru(CN)_6$, $Mo(CN)_8$, $W(CN)_8$, $Fe_{(aq)}$, $Eu_{(aq)}$ and $V_{(aq)}$.

Outer-sphere redox probes are termed *surface insensitive* such that k^o is not influenced by the surface oxygen-carbon ratio, surface state/cleanliness in terms of a surface coating of a monolayer film of uncharged adsorbates, or specific adsorption to surface groups/sites. There is no chemical interaction or catalytic mechanism involving interaction (i.e. an adsorption step) with the surface or a surface group in this case the electrode merely serves as a source (or sink) of electrons and as such outer-sphere systems are sensitive primarily to the electronic structure due to the density of electronic states near the Fermi level.

On the other hand, inner-sphere redox mediators are termed *surface sensitive* in that the k^o is strongly influenced by the state of the electrode surface (surface chemistry and microstructure) via specific electro-catalytic interactions that are inhibited significantly if the surface is obscured by adsorbates (or impurities). Such interactions can also depend strongly on the presence (or absence) of specific oxygenated species which give rise to either beneficial or detrimental electrochemical effects [10, 11]. In this case systems are more largely affected by surface state/structure and/or require a specific surface interaction, being catalysed (or inhibited) by specific interactions with surface functional groups (adsorption sites) rather than the DOS as such systems generally have high reorganisation energies [10, 11].

The observation of differing responses when using varied inner- and outer-sphere redox probes allows insights to be deduced regarding the state of the surface structure of the electrode material in question. McCreery [10, 11] has provided a “road map” for commonly utilised redox probes, as shown in Fig. 4.6, which allows researchers to clarify from experimental observations the redox systems and how they are affected.

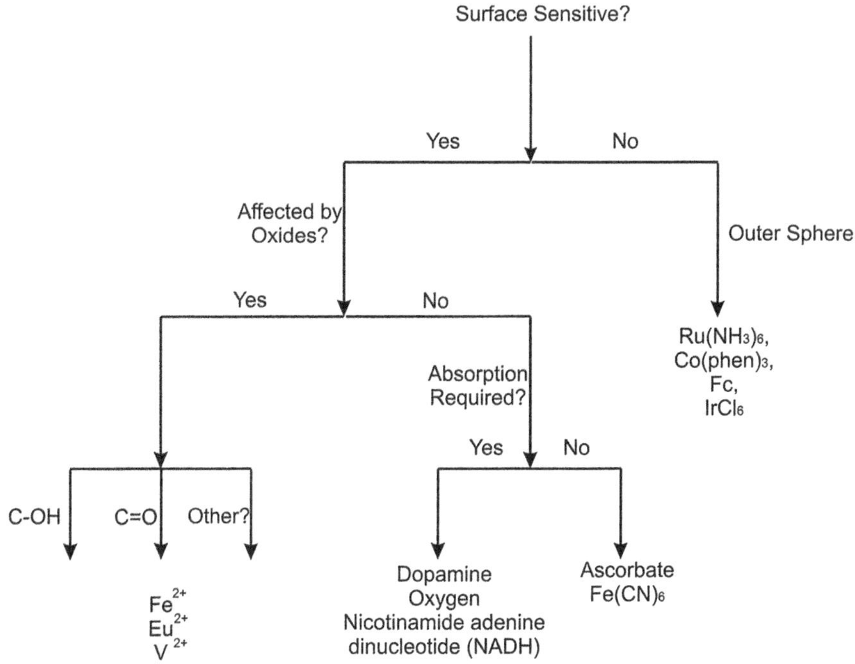

Fig. 4.6 Classification of redox systems according to their kinetic sensitivity to particular surface modifications on carbon electrodes. Adapted from Ref. [9]

4.3 Changing the Electrode Geometry: Macro to Micro

As mentioned in Chap. 1, macro- and micro-electrodes can be fabricated via screen-printing. At a macroelectrode, electrolysis of *A* occurs across the entire electrode surface such that the diffusion of *A* to the electrode or *B* from the electrode surface is termed planar, and the current response is typically described as 'diffusion limited', giving rise to an asymmetric peak as shown in Fig. 4.7a. At the edge of the macroelectrode, where the electrode substrate meets the insulting material defining the electrode area, diffusion to or from the edge of the electrode is effectively to a point. Therefore, the flux, *j*, and the rate of mass transport are larger at the edge and as such diffusion becomes convergent. This is termed an 'edge effect' which is negligible at a macroelectrode since the contribution of convergent diffusion to the edges of the macroelectrode is inundated by that of planar diffusion to the entire electrode area.

As the electrode size is reduced from macro to micro, or even smaller to that of nano, convergent diffusion to the edges of the electrode becomes significant. In this regime a change in the observed voltammetric profile is observed which results in the loss of the peak shaped response, as evident in Fig. 4.7b with that of a sigmoidal voltammogram. The effect of convergent diffusion has the benefit of improvements

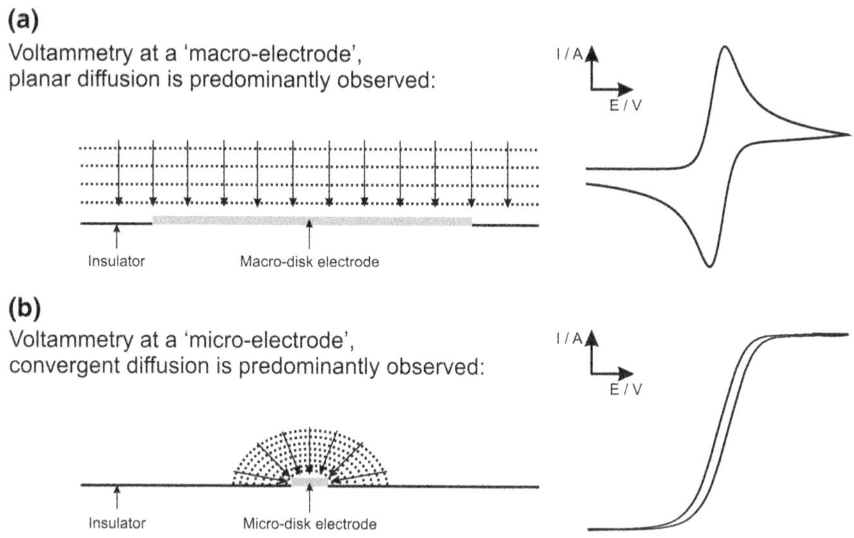

Fig. 4.7 The unique differences between the cyclic voltammetric signatures observed at a macroelectrode (**a**) compared to a microelectrode (**b**)

Fig. 4.8 Steady state voltammogram for a reversible process observed at a microelectrode

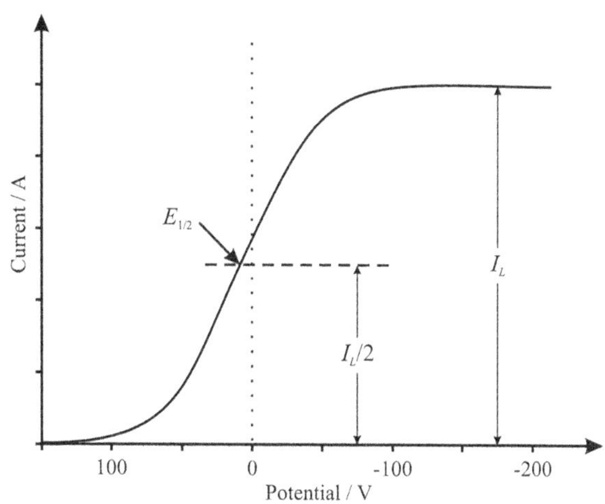

in mass transport such that the current density is greater than at a macroelectrode under planar diffusion (Fig. 4.8).

For a reversible electrode reaction at a microelectrode, as shown in Fig. 4.7b, where $E_{1/2}$ is the half-wave potential, the following equation describes the expected voltammetric shape:

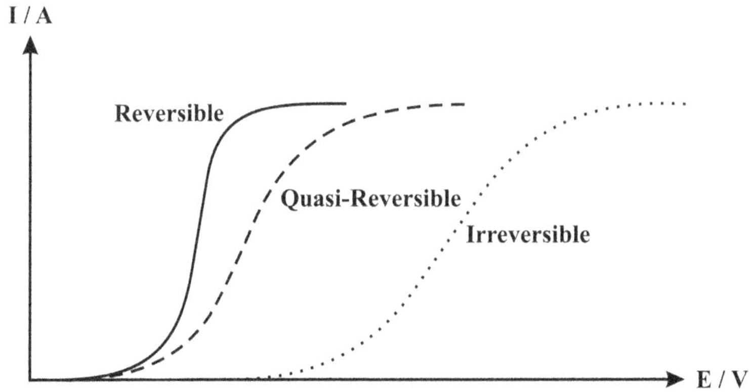

Fig. 4.9 How a steady-state voltammogram is shaped by electrochemical (heterogeneous) kinetics

$$E = E_{1/2}^{rev} + \frac{RT}{nF} In \frac{I_L - I}{I_L}$$
$$\text{where } E_{1/2}^{rev} = E^{0'} + \frac{RT}{nF} In \frac{D_R^{1/2}}{D_0^{1/2}} \tag{4.15}$$

since the ratio of diffusion coefficient is nearly equal, $E_{1/2}$ is a good approximation for $E^{0'}$ for a reversible couple.

When a plot of E against $In\frac{I_L - I}{I_L}$ is constructed, a linear response should be observed with a gradient equal to RT/nF and an intercept of $E_{1/2}^{rev}$ if the wave-shape corresponds to a reversible process. The effect of different electrochemical kinetics is shown for the case of a microelectrode in Fig. 4.9, where the voltammogram is shifted as the electron transfer becomes slower since a greater 'overpotential' is needed to overcome the kinetic barrier. In this case, Eq. (4.15) becomes: $E = E_{1/2}^{irr} + \frac{RT}{\alpha nF} In \frac{I_L - I}{I_L}$. Thus a plot of E against $In\frac{I_L - I}{I_L}$ gives rise to a gradient of $\frac{RT}{\alpha nF}$ and an intercept of $E_{1/2}^{irr}$. To determine between reversible, quasi-reversible and irreversible, a useful approach is the Tomeš criteria [12]; see Ref. [13] for a full overview of the various diagnostic approaches.

Last, Fig. 4.9 shows the different microelectrode geometries that can be readily encountered in electrochemistry. For an elegant overview of microelectrodes and their benefits and applications, readers are directed to Ref. [14].

In the above part of this springer brief we have considered cyclic voltammetry and its derivatives. Another technique that is worthy of mention, and that can be used to study screen-printed electrodes, is chronoamperometry. This technique is commonly used either as a *single potential step*, in which only the current resulting from the forward step (as described above) is recorded, or *double potential step*, in which the potential is returned to a final value following a time period, usually designated as τ, at the step potential. The electrochemical technique of chronoamperometry involves stepping the potential applied to the working electrode, where initially it is

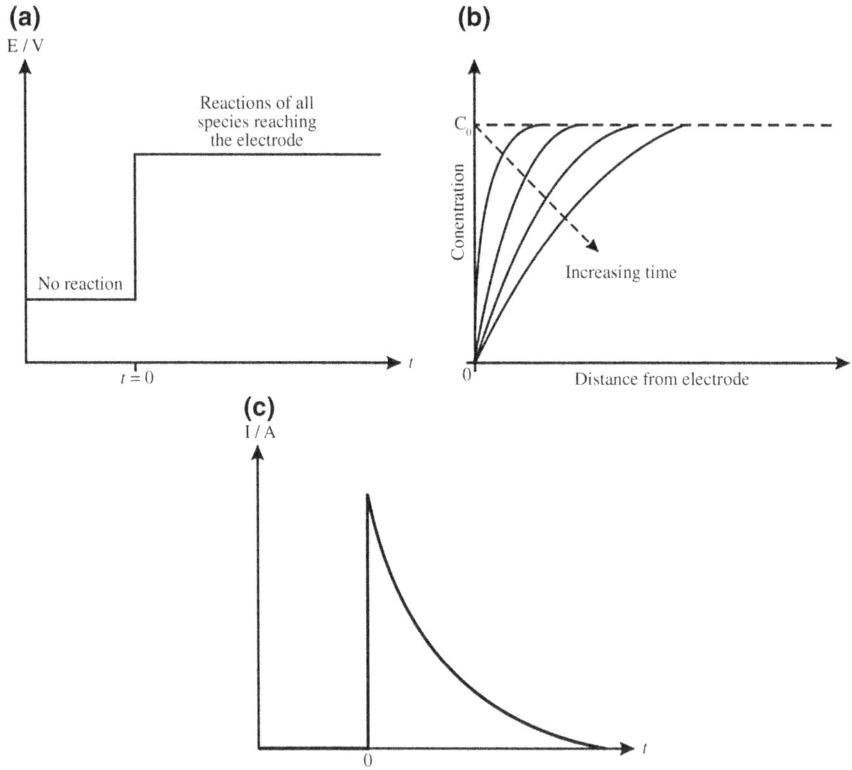

Fig. 4.10 Chronoamperometric experiment: **a** potential-time waveform; **b** change of concentration gradient; **c** resulting current-time response

held at a value at which no Faradaic reactions occur before jumping to a potential at which the surface concentration of the electroactive species is zero Fig. 4.10a, where the resulting current-time dependence is recorded Fig. 4.10c.

The mass transport process throughout this process is solely governed by diffusion, and as such the current-time curve reflects the change in concentration at the electrodes surface. This involves the continuing growth of the diffusion layer associated with the depletion of reactant, thus a decrease in the concentration gradient is observed as time progresses Fig. 4.10b. An example of single potential step chronoamperometry is shown in Fig. 4.11 for the case of an Osmium complex modified electrode and also shown in the inserts are the effect of concentration of epinephrine which is electro-catalysed undergoing an EC′ process (see Sect. 2.5); note that this is the basis of commercially available glucose biosensors.

The most useful equation in chronoamperometry is the Cottrell equation, which describes the observed current (planar electrode of infinite size) at any time following a large forward potential step in a reversible redox reaction (or to large overpotential) as a function of $t^{-1/2}$.

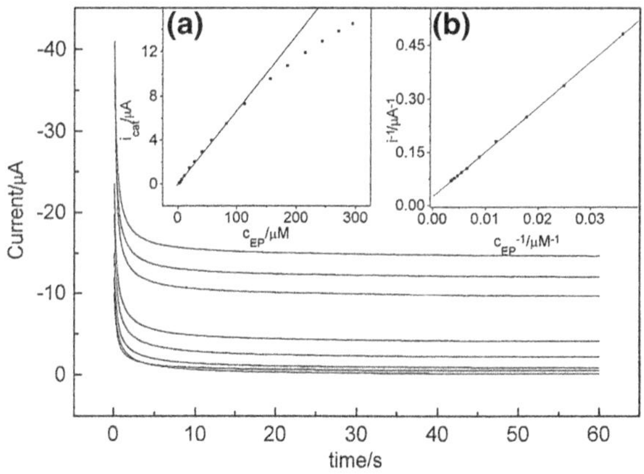

Fig. 4.11 Chronoamperometric curves with a potential step from 0 to +0.4 V at Os-(PVP) 10/Nafion modified electrode in pH 6.9 PBS containing 0, 6.5, 11, 28, 56, 156, 215 and 294 μM (from *bottom* to *top*). *Inset*: (**a**) plot of catalytic current versus epinephrine concentration; (**b**) data analysis of catalytic current versus. epinephrine concentration. Reproduced from Ref. [15] with permission from Elsevier

$$I_L(t) = nFAD^{1/2}C(\pi t)^{-1/2} \tag{4.16}$$

where n = stoichiometric number of electrons involved in the reaction, F is the Faraday's constant, A is the electrode area, C is the concentration of electroactive species and D is the diffusion coefficient. The current due to double layer charging also contributes following a potential step but decays as a function of $1/t$ and is only significant during the initial period, typically a few milliseconds following the potential step (Fig. 4.12).

4.4 Surface Characterisation

It is vital when utilising such screen-printed architectures or any electrochemical setup that the visual and surface characterisation of such system is analysed, therefore informing the user on the true surface of the electrode. A common technique for this is scanning electron microscopy (SEM) which allows the user to analysis the surface and produce images of the areas that could be potentially interesting. Shown within Fig. 4.13 is a typical SEM image of the screen-printed electrodes, it is visible that there are areas of large comprising of a range of carbon/graphitic material, solvents and polymeric binders. The latter are electrochemically inactive [17] and will change the true electrochemically active area. On the other hand, due to the way the inks are formulated and depending on the quality of the print, the electrode might be significantly rough. As such the fabrication and formulation of the inks and the quality of the print, the electrode might be

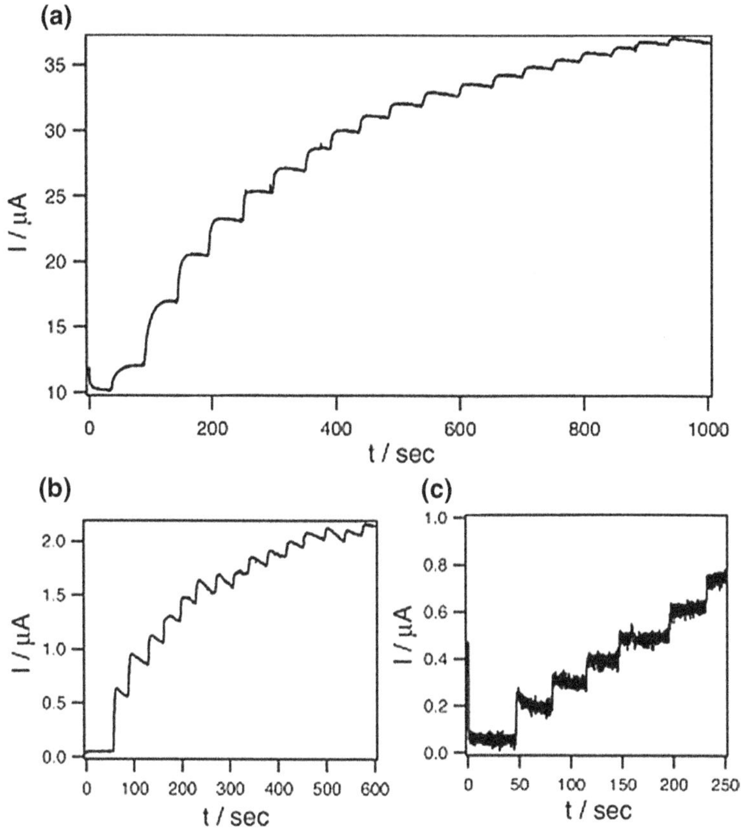

Fig. 4.12 Current–time curve obtained for ITO/AuNP/cyt c electrode upon successive addition of 20 μL aliquots of 200 mM H_2O_2 to 5 mL stirred 10 mM HEPES buffer at pH 7 with an applied potential of −0.1 V under nitrogen atmosphere; chronoamperometric curve obtained for (**b**) cyt c/ITO and (**c**) AuNP/ITO obtained by the addition of 20 μL aliquots of 200 mM H_2O_2 in 5 mL stirred solution of 10 mM HEPES buffer at the potential of −0.1 V under nitrogen atmosphere. Reproduced from Ref. [16] with permission from Elsevier

significantly 'rough'. McCreery et al. [18] have shown an elegant schematic of the effect upon the electrochemical properties of the system due to the diffusion layer when a rough surface is introduced [18]. It is clear in Fig. 4.14a that the diffusion layer is larger than surface roughness and in Fig. 4.14b the opposite occurs, creating a response mimicking a larger surface area.

Thus in most cases, the diffusion layer is thicker than that of the electrode diameter and its surface roughness and the (diffusional limited electroanalytical response is not dramatically affected by the surface roughness. This of course changes when fast voltammetry is undertaken. An issue with using voltammetry is that the electrode has to be used and cannot be used in the internal approach. Thus,

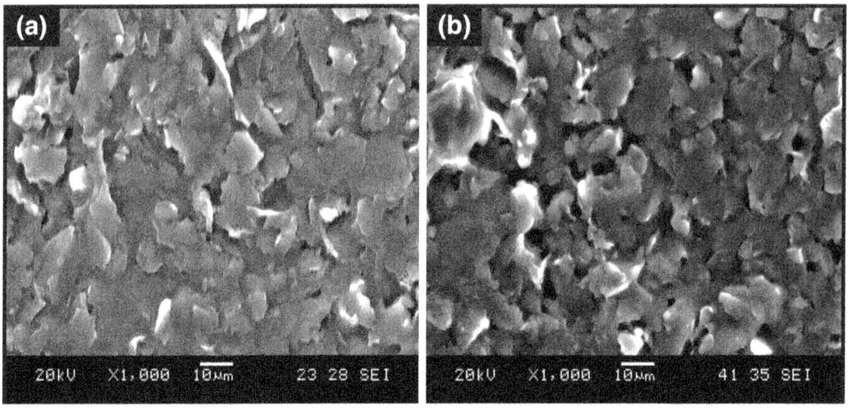

Fig. 4.13 Typical SEM images of a graphitic screen-printed electrode

Fig. 4.14 Comparison of electrode roughness against the diffusion layer, δ, adapted from Ref. [18]

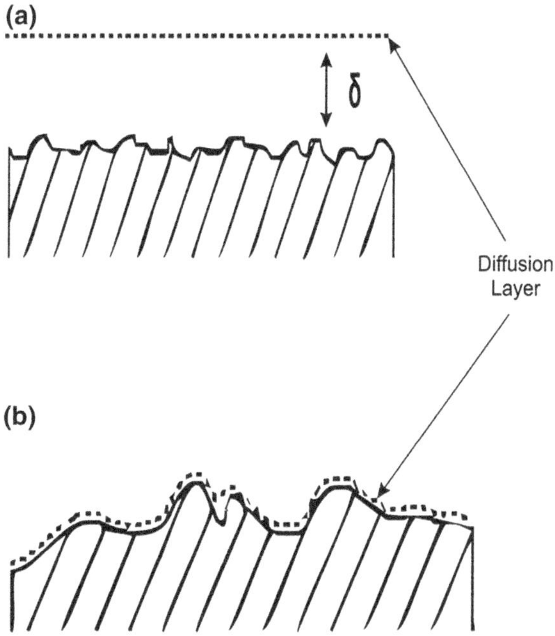

another method is to sample a random number from a batch to try and determine whether the overall batch conforms to set standards.

In addition to the aforementioned methods, an optical approach can be carried out to map the electrode surface; this can be used to define the diameter and circumference of the electrode; allowing the fabricator to reject sensors that do not meet the criteria expected. However this will not give an indication of the electrode roughness. A method to measure the surface roughness which can potentially be used in QC/QA analysis of the fabricated electrodes is white light profilometry. From this technique a surface roughness profile can be achieved measuring the surface topography of the screen-printed electrode to determine a roughness factor (R_F), and are represented in a surface map of the electrode.

The surface profile maps can be analysed using a Matlab script based on the following equation:

$$R_F = \frac{\sum_{k=0}^{M-2} \sum_{i=0}^{N-2} A_{kl}}{(M-1)(N-1)\delta x \delta y} \tag{4.17}$$

where M and N are the total number of points in the x and y directions respectively, x and y are the distances between the points in the x and y directions, and where:

$$A_{kl} = \begin{aligned} &\tfrac{1}{4}\left(\sqrt{\delta y^2 + (z(x_k, y_i) - z(x_k, y_{i+1}))^2} + \sqrt{\delta y^2 + (z(x_{k+1}, y_i) - z(x_{k+1}, y_{i+1}))^2}\right) \\ &+ \left(\sqrt{\delta x^2 + (z(x_k, y_i) - z(x_{k+1}, y_i))^2} + \sqrt{\delta y^2 + (z(x_k, y_{i+1}) - z(x_{k+1}, y_{i+1}))^2}\right) \end{aligned} \tag{4.18}$$

with z being the height above the surface at a coordinate (x, y). Although similar to the surface area ratio (S_{dr}) typically used for surface topology measurements [19], the equation has been modified to provide the ratio of interfacial area to the area of the projected horizontal plane, rather than the increment of the interfacial surface area to the projected horizontal plane. This modification has been made as the roughness factor and is described as the ratio of the active surface area to the substrate geometric surface area [20]. This approach also allows for the analysis of the thickness of a deposit from the resulting screen-print, as utilising different inks will give varying thicknesses. It is noted that if the roughness factor starts to change significantly through a batch it could be the case that the ink requires a greater agitation prior to printing.

4.5 Quality Control of Screen-Printed Electrodes

In summary the quality control of these systems can include:

- A visual inspection for print quality
- In-process inspection checks such as the resistance measurement of the dried printed structure, image stretch and layer-to-layer registration measurements
- Electrochemical examination (cyclic voltammetry) across printed card, row to row and through batch measuring *Ip*, *Ep* and k^0

- Microscopic measurement of the diameter and/or geometric area of the dried ink deposit upon the screen-printed electrode using a coordinate measuring machine
- The measurement of deposition thickness via white light profilometry. A combination of the above can be used to determine the quality of a batch of screen-printed electrodes.

A combination of the above can be used to determine the quality of a batch of screen-printed electrodes.

4.6 Quantifying the Composition of the Screen-Printed Reference Electrode

The silver/silver chloride reference electrode is printed from an appropriately formulated ink containing (amongst other components) significant amounts of solid silver and silver chloride. A range of silver: silver chloride ratios are available from commercial ink suppliers such as: 40:60; 50:50; 60:40, 70:30; 80:20; 85:15 and 90:10. We show below what effect this has on the voltammetry but note that as the amount of silver increases, the price of the ink substantially increases due to the high cost of silver metal.

The electrochemical process at the reference electrode is as follows:

$$AgCl(s) + e^- \rightleftharpoons Ag(s) + Cl^-(aq) \tag{4.18}$$

The appropriate Nernst equation for this process is described by Eq. (4.19):

$$E = E^{0'} + \left(\frac{RT}{nF}\right) In\left[\frac{[AgCl]}{[Ag][Cl^-]}\right] \tag{4.19}$$

where, $E^{0'}$ is the formal potential. Equation (4.19) can be rewritten as:

$$\begin{gathered} E = E^{0'} + \left(\frac{RT}{nF}\right) In\left[\frac{k}{[Cl^-]}\right] \\ \text{where} \quad k = \frac{[AgCl]}{[Ag]} \end{gathered} \tag{4.20}$$

Through investing and separating the logarithmic terms and conversion of base 10 gives [20]:

$$E = E^{0'} - 2.303\left(\frac{RT}{nF}\right)[\log[Cl^-] - \log(k)] \tag{4.21}$$

Assuming the experiment is run at 298 K gives:

$$E = E^{0'} - 0.0592\log[Cl^-] + 0.0592\log(k) \quad (4.22)$$

Hence one can clearly see that the electrode potential of this reference electrode decreases by ~59 mV for every decade change in chloride ion concentration; not surprisingly one will see that when electroanalytical measurements are reported in the literature that the chloride ion concentration is fixed otherwise the potential will shift accordingly. What is also insightful is that if the ratio (k) of silver:silver chloride in the reference electrode changes then the measured potential will also change. This could happen unexpectedly during printing a large batch that a new reference electrode ink is opened where the ratio of silver:silver chloride is different to that used previously. This would mean that the measured potential would change during the batch and would greatly affect the sensor performance, particularly is the measurement protocol is based upon chronoamperometry. Last, through the use of Eq. (4.20) the ratio (k) can be experimentally determined to ensure that the ink received from the manufacturer is as claimed.

4.7 How to Electrically Wire to Your Fabricated Screen-Printed Architectures

Researchers can clearly see the allure of utilising screen-printed electrodes and newcomers to the field often fail to obtain reproducible and useful electrochemical measurements which are typically blamed upon the screen-printed electrodes themselves. Others lay the problem with connection to the electrode being difficult, typically utilising crocodile clips and thus obtain electrical noise in the electrochemical signals resulting in unreliable limits the detection, in terms of the analytical sensitivity and limit of detection, towards the target analytical probe. Both such scenarios inevitably result in researchers abandoning the use of screen-printed electrodes. Motivated by these problems, in this technical note the often overlooked factor of screen-printed electrode experimental setups is explored between the use of an edge connector and the crocodile clips both used in the electrical wiring of the SPEs to the potentiostat/electronics (Fig. 4.15).

Fig. 4.15 Optical image of a SPE working electrode

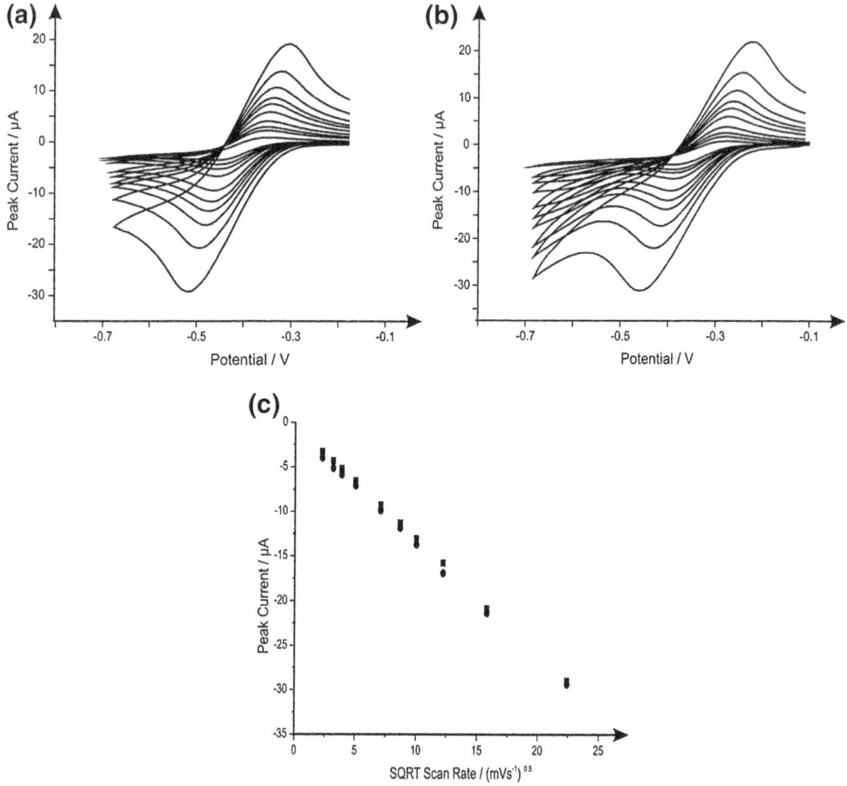

Fig. 4.16 Typical cyclic voltammograms using both an edge connector (**a**) and a crocodile clip setup (**b**) recorded in 1 mM $[Ru(NH_3)_6]^{2+/3+}$ in pH 7.4 PBS/0.1 M KCl. Also shown are plots of peak height versus square root of scan rate (**c**) for the edge connector and crocodile clip setup (*squares* and *circles* respectively). Reproduced from Ref. [22]

Electrochemical comparison between the two setups of the traditional approach of crocodile clip connection with that of the edge connector, the employment of crocodile clips for electrochemical analysis within a laboratory environment *can produce* near identical responses with that of an edge connector. When one diligently connects the crocodile clips, as shown in Fig. 4.16. Appropriate separation of the crocodile *must* be applied to create these responses, shown in Fig. 4.17a and not like that in Fig. 4.17b, the latter gives rise to useless voltammetric signatures while the former gives near perfect and expected voltammetric responses; some researchers obtain such results as presented in Fig. 4.17c and unfortunately blame the electrode itself without further explanation.

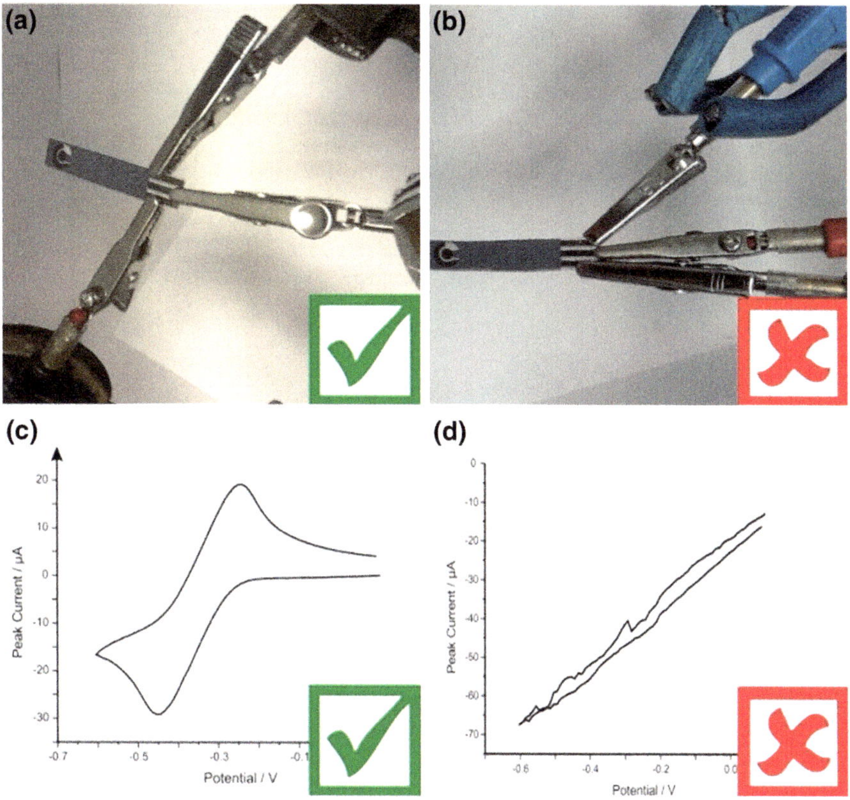

Fig. 4.17 Images showing the correct (**a**) and incorrect (**b**) way to connect the screen-printed electrodes/sensors to the potentiostat/electrodes utilising crocodile clips, with their corresponding cyclic voltammograms (**c** and **d** respectively) recorded in 1 mM $[Ru(NH_3)_6]^{2+/3+}$ in pH 7.4 PBS/0.1 M KCl. Reproduced from Ref. [22]

References

1. Bard AJ, Faulkner LR (2001) Electrochemical methods, Second edn. Wiley-VCH, New York
2. Compton RG, Banks CE (2011) Understanding voltammetry, Second edn. Imperial College Press, London
3. Wang J (2000) Analytical electrochemistry, Wiley-VCH. Second edn, New York
4. Brownson DAC, Banks C E (2014) Handbook of graphene electrochemistry, Springer
5. Matsuda H, Ayabe Y (1955) Z Elektrochem 59:494–503
6. Nicholson RS (1965) Anal Chem 37:1351–1355
7. Klingler RJ, Kochi JK (1981) J Phys Chem 85:1731–1741
8. Lavagnini I, Antiochia R, Magno F (2004) Electroanalysis 16:505–506
9. Cline KK, McDermott MT, McCreery RL (1994) J Phys Chem 98:5314–5319
10. McCreery RL (2008) Chem Rev 108:2646–2687
11. Chen P, McCreery RL (1996) Anal Chem 68:3958–3965
12. Tomeš J (1937) Coll Czech Chem Comm 9:12–21

13. Bard AJ, Faulkner LR (2001) Electrochemical methods: fundamentals and applications, 2nd edn. John Wiley & Sons Inc, United States of America
14. Stulik K, Amatore C, Holub K, Marecek V, Kutner W (2000) Pure Appl Chem 72:1483–1492
15. Ni J-A, Ju H-X, Chen H-Y, Leech D (1999) Anal Chim Acta 378:151–157
16. Yagati AK, Lee T, Min J, Choi J-W (2012) Colloids Surf B 92:161–167
17. Choudhry NA, Kadara RO, Banks CE (2010) Phys Chem Chem Phys 12:2285–2287
18. McCreery RL, Cline KK (1996) Laboratory techniques in electroanalytical chemistry, Second edn. Marcel Dekker Ltd, pp. 293–331
19. Patois T, Lakard B, Monney S, Roizard X, Fievet P (2011) Synth Met 161:2498–2505
20. Benck JD, Chen Z, Kuritzky LY, Forman AJ, Jaramillo TF (2012) ACS Catal 2:1916–1923
21. Cranny A, Harris NR, Nie M, Wharton JA, Wood RJK, Stokes KR (2011) Sens Actuators, A 169:288–294
22. Galdino FE, Foster CW, Bonacin JA, Banks CE (2015) Anal Methods 7:1208–1214

Printed by Printforce, the Netherlands